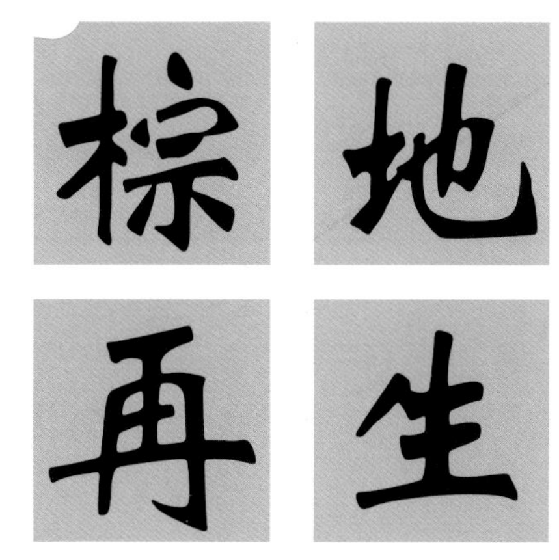

Brownfield Regeneration

2020 河北省第四届（邯郸）园林博览会
设计实践与技术应用

Design and Technique Application of the Fourth Hebei (Handan) Garden Expo, 2020

俞孔坚 / 刘向军 / 张媛 / 王书芬 著
Yu Kongjian / Liu Xiangjun / Zhang Yuan / Wang Shufen

中国建筑工业出版社

图书在版编目（CIP）数据

棕地再生：2020河北省第四届（邯郸）园林博览会设计实践与技术应用 = Brownfield Regeneration Design and Technique Application of the Fourth Hebei (Handan) Garden Expo，2020／俞孔坚等著．—北京：中国建筑工业出版社，2021.7
ISBN 978-7-112-26330-1

Ⅰ．①棕… Ⅱ．①俞… Ⅲ．①园林设计－研究－邯郸 Ⅳ．TU986.2

中国版本图书馆CIP数据核字（2021）第135974号

责任编辑：郑淮兵　王晓迪
责任校对：王　烨

棕地再生　2020河北省第四届（邯郸）园林博览会设计实践与技术应用
Brownfield Regeneration Design and Technique Application of the Fourth Hebei (Handan) Garden Expo, 2020
俞孔坚　刘向军　张媛　王书芬　著
*
中国建筑工业出版社出版、发行（北京海淀三里河路9号）
各地新华书店、建筑书店经销
北京方舟正佳图文设计有限公司制版
北京富诚彩色印刷有限公司印刷
*
开本：787毫米×1092毫米　1／12　印张：23⅓　字数：702千字
2021年7月第一版　2021年7月第一次印刷
定价：**268.00**元
ISBN 978-7-112-26330-1
（37692）

前言

2000 年，在广东省中山市原有造船厂遗址，我们首次尝试城市土地的再利用途径，设计了反映工业化时代文化特色的公共休闲场所——岐江公园，开创了国内工业遗产设计的先河；2010 年，在原为钢铁厂（浦东钢铁集团）和船舶修理厂的上海后滩棕地遗址，利用世博会契机，我们将其改造为黄浦江边的公共绿地空间——后滩公园，再次通过工业遗产的成功再造为城市带来了可持续的发展生机；如今，2020 年，面对邯郸市复兴区面积达 3km^2 的工业棕地——邯钢矿渣旧址，我们应用生态修复综合措施将其改造为具有山水格局的森林中的绿色的园博园，这一项目的实施建设可以看作是对工业遗产再利用理念最广泛的一次社会普及。

2018 年 11 月底，我们到场地进行初次踏勘，见证了钢城遗址工业废弃地的原貌：遍布矿渣、矿泥；偶然见到的相对平坦的地方是硬化的工厂厂房原址；水渣场周边废弃矿渣堆积如山，底部渗出白色液体流成“小溪”，汇入有“邯郸母亲河”之称的沁河；而邯郸西部的代表湿地——西湖，由于附近村庄就近排污以及驳岸年久失修，似一颗蒙尘的珍珠遗落在复兴区的土地上。

经过对 4200 多亩（约 280hm^2）规划现场的深度勘查以及对邯郸 3000 年文化的挖掘，在首席设计师俞孔坚老师的带领下，我们于 2018 年 12 月底提交出一套具有现代特色的园博会方案。省级领导和具有办会经验的不同专家组对方案多次进行审核，我们消化吸收了相关意见并对方案进行了修改提升，最终确定了以“山水邯郸，绿色复兴”为主题的生态规划设计方案。此时距预定开园时间已不足 18 个月，我们马不停蹄对项目进行方案深化并协同业主提取场地内所有市政条件及限制性因素，如：规划范围内两处村庄有市政设施从场地内穿过，设计需保留；“先有涧沟，后有邯郸城”之称的涧沟村周边散落着大大小小十几处国家级汉墓遗址的保护范围；场地最西侧 50m 处为南水北调工程穿越区，需要 1000m 保护范围。此外，区域内保留村庄 630 多亩（约 42hm^2）、基本农田 300 多亩（约 20hm^2）、现状水库 300 多亩（约 20hm^2），去除保护文物以及建设单位重视的保护项目等，核心建设区约 1800 亩（约 120hm^2），基本为地质复杂的矿渣废弃地。方案深化阶段，我们又分专项分节点向省市领导及专家汇报并进行单项审核，在方案阶段确保项目的可实施性。据不完全统计，至项目完成，各种汇报总次数达 60 次以上，这在我们之前的项目设计中是绝无仅有的，这也充分体现了项目本身的综合性和复杂性。

因此项目采用 EPC 合作模式，设计团队从施工最初阶段至开园，乃至整个运营期，全程驻场跟踪配合，牵头参与项目交底，编制部分施工组织计划，跟业主随时汇报沟通项目进程和各种突发问题并及时解决各种诉求，跟踪把控施工过程中每一环节的实施效果，从场地清表、永临结合水电和道路的布置、水电管线预埋到地形塑造；从乔木栽植、支护方式、修剪技术的指导到地面铺装的接缝处理；从垒砌挡墙的细节到耐候钢板的喷漆，每一个细节处理、每一处施工场地都有设计师们的身影，我们指导施工、参与施工，甚至自己动手去铺砌卵石、栽植花草，只为呈现更好的效果。在项目工程步入正轨之际，突遇全民抗疫的特殊事件，面对疫情考验，邯郸市复兴区领导带领整个团队，力争抗疫施工两不误！指挥部办公室启动随时待命的工作模式，疫情导致施工人员队伍不能正常到位，外省道路不通，材料进场困难……一系列问题的发生，都为开园效果留下些许遗憾。特别是受制于土建工程的完工时间，地被花卉种植时间紧张，我们不得不临时整合优化新的应急方案。依据当地材料信息并结合项目现场进度及各种情况快速整合优化针对开园的可行性地被种植方案，设计师们分地块进行地毯拉网式指导施工现场效果，做种植示范，俞孔坚老师也多次亲临现场指导，以期达到理想的建成效果，最大限度实现我们的设计初衷及理念！此番回首，百感交集，在所有人的共同坚持之下，邯郸园林博览会于 2020 年 9 月 16 日顺利开园！至 2020 年 11 月 28 日园博园闭园，共接待了近 70 万游客。园博园的精彩设计及建成效果受到了来自各省市级领导及广大游客的高度评价和赞赏。

本次邯郸园博会项目的总体设计，是一次城市生态修复的成功案例，也是国内完全采用现代园林设计手法完成的首个园博会项目；是一次 EPC 项目协作的经验积累，也是对棕地再生设计的一次全面归纳。在本书中我们总结出规划、建筑、景观实践的做法，以期为对未来同类项目的技术推广起到示范作用。

土人设计

2021 年 1 月 8 日

Introduction

In 2000, at the former shipyard site in Zhongshan City, Guangdong Province, we tried for the first time the reuse of urban land which resulted in Qijiang Park. As a public leisure place reflecting the cultural characteristics of the industrialized era, it set a precedent for China's industrial heritage pavilion design. In 2010, we took the opportunity of the World Expo and transformed the Shanghai Houtan brownfield site, which was formerly occupied by a steel plant (Pudong Steel Group) and a ship repair plant, into a public green space by the Huangpu River—Houtan Park. Through the reconstruction of an industrial heritage pavilion, we once again brought sustainable development vitality to the city. Now in 2020, in front of the industrial brownfield covering an area of 3 square kilometers in Fuxing District, Handan City, which used to be the slag site of Handan Steel, we applied ecological restoration measures to transform it into a green garden expo park with a landscape pattern. The construction of this project can be regarded as one of the most extensive social popularization works of industrial heritage pavilion reuse.

At the end of November 2018, we made the initial survey and witnessed the original appearance of the industrial wasteland of the steel plant, where slag and sludge were all over the place. The occasional flat area was the hardened site of the former factory building. At the bottom of the abandoned slag pile, white liquid seeped out and flowed into the Qin River, the mother river of Handan. The featured wetland in the west of Handan, the Xihu Wetland, suffered from the discharge of the nearby villages and the revetment in disrepair.

After an in-depth survey of the more-than-4200-mu (280 hectares) planning site and Handan's three-thousand-year culture, led by the chief designer Prof. Yu Kongjian, we submitted a set of plan with modern features at the end of December 2018. Provincial officials and expert groups experienced in expo holding reviewed the plan for multiple rounds. We revised the plan according to their inputs, and finalized an ecological planning and design with the theme of "Natural Handan, Green Revival". At the time, it was less than 18 months till the scheduled opening day. We rapidly carried out the design development and extracted restrictive factors in the site, such as, municipal facilities passing through two villages which needed to be maintained. Jiangou Village, known as the origin of the Handan City, is surrounded by dozens of scattered protection areas of national-level Han tomb sites. The westernmost 50 meters of the site is crossed by the South-to-North Water Diversion Project, which requires 1,000m of protection range. In addition, more than 630 mu (42 hectares) of village-level farmland, 300 mu (20 hectares) of basic farmland, and 300 mu (20 hectares) of current reservoirs are retained in the site. The projects of cultural relics protection and the ones that the contractor deemed important were removed. The core construction area is about 1800 mu (120 hectares), which is basically a slag waste land with complex geology. In the design development stage, we reported to the government and experts in sub-nodes and conducted individual audits to ensure the feasibility of the plan. According to incomplete statistics, more than 60 reports were conducted before we reached the completion of the project, which was unprecedented in our previous projects—which also reflects the comprehensiveness and complexity of the project itself.

Therefore, EPC cooperation model was adopted in the project. Through the entire period of design, opening and operation of the project, the design team stayed on site and took the lead in design interpretation, prepared part of the construction plan, and timely reported to the owner to communicate the progress of the project. In addition, the designers also dealt with emergencies, met various demands, as well as tracked and controlled the implementation effect of every section of the construction, including site clearing, integration of permanent and temporary layout of hydropower and roads, pre–burying hydropower pipelines, terrain shaping, tree planting and support methods, pruning technology guiding, paving joint treatment, retaining wall details, spray paint of the weather–resistant steel plate, etc. Our designers were present at every detail processing and every construction site. We guided and participated in the construction. We laid the pebbles and planted the trees and flowers for the sake of better result effect. Just when the project was on track, the whole country ran into the special event of the COVID–19 epidemic. Facing the challenge, the government leaders of Fuxing District led the entire team to neglect neither the epidemic fighting nor the construction! The headquarters office started an on–call work mode. With the inter–provincial roads blocked, the construction staff and construction materials faced difficulties in entering the site. A series of problems occurred, leaving us regrets for the effect of the opening of the park. In particular, constrained by the completion time of the civil works, there was little time left for flowers and ground covers planting, forcing us to temporarily make new emergency plans. According to the results of local material information and combined with the project progress, we rapidly optimized the feasible ground cover planting plan for the opening of the park. The designers were distributed to different patches and guided the construction by personally demonstrating the planting. Even Prof. Yu Kongjian visited the site multiple time in order to achieve the ideal construction effect, so as to realize our original design intention to the maximum extent! Now looking back, with mixed feelings, thanks to the common hard work of all members of the team, the Handan Garden Expo Park was successfully opened on September 16, 2020! As of November 28, 2020, nearly 700,000 visitors had been accepted. The wonderful design and construction of the Garden Expo Park has been highly praised and appreciated by the broad masses.

The overall design of the Handan Garden Expo Park is a successful case of urban ecological restoration. It is also the first domestic garden expo project completed using modern garden techniques. It is an accumulation of experience in EPC project collaboration, as well as a comprehensive summary of brownfield regeneration design. In this book, we summarize the practices of planning, architecture and landscape of the project, with a view to playing a demonstrative role in the promotion of the techniques for similar projects in future.

Turenscape
January 8, 2021

本书主要贡献者

设计单位：北京土人城市规划设计股份有限公司
首席设计师：俞孔坚
项目总负责人：刘向军
项目总协调：张志文、边文光、刘德华
景观专业负责人：王书芬、张要刚
建筑专业负责人：张媛、金磊
设备专业负责人：陈娆、李昊、鲁昂
现场负责人：郑亚凯、郑军彦
设计团队：李江涛、孟庆芳、秦玥、张帆、徐嘉、阮甜子、朱青、张雪松、刘杰、何选宁、方化武、李子轩、刘伟、宋菲、郑姝伟、罗京晶、雷嘉、陈萌萌、冯阳、周一男、史育玉、孙华、陈素波、彭英豪、李鑫、温明、何俞轩、刘勇、刘嘉琪、徐艳玲、刘树梅、赵兴雅、李东辉、吴昊、王磊、屈秀娟、宗丽娜等

设计委托与建设方：邯郸市复兴城市和交通建设投资有限公司
水生态研究合作单位：中国科学院生态环境研究中心
建筑结构咨询合作单位：河北建筑设计研究院有限责任公司

特别感谢：
邯郸市委市政府
邯郸市复兴区委区政府
邯郸市复兴区筹备河北省第四届园博会工作推进指挥部：潘利军（复兴区区委书记）、李少锋（复兴区区委副书记、政府区长）、白建功（复兴区区政协主席）、汪守君（复兴区政府副区长）、李忠民（邯郸市复兴城市和交通建设投资有限公司董事长）
邯郸市自然资源和规划局
邯郸市城管局
专家审核组的曹南燕（住房和城乡建设部风景园林专家委员会委员）、包满珠（华中农业大学园艺林学学院院长）、郑占峰（邢台市自然资源和规划局）、张金江（河北工程大学副校长）、唐学山（北京林业大学教授、博士生导师）、赵世伟（北京植物园原园长）、张树林（北京市园林绿化局原副局长）、任春秀（辽宁省住房和城乡建设厅原总风景园林师）、张晓鸥（江苏省住房和城乡建设厅原调研员）、李秀云（唐山市园林绿化管理局副局长）

目录
CONTENTS

前言 3
INTRODUCTION

概要 11
ESSENTIALS

1 项目概况 12
PROJECT OVERVIEW

1.1 项目背景 14
PROJECT BACKGROUND

1.2 文化脉络 18
CULTURAL TRADITIONS

1.3 外部交通 20
TRANSPORT ACCESSIBILITY

1.4 场地认知 22
SITE COGNITION

1.4.1 自然条件 24
NATURAL CONDITIONS

1.4.2 人文资源 30
CULTURAL RESOURCES

1.4.3 限制因素 32
CONSTRAINT CONDITIONS

2 挑战与策略 35
CHALLENGES AND STRATEGIES

2.1 建设挑战 36
CONSTRUCTION CHALLENGES

2.2 设计主题与目标 40
DESIGN THEME AND GOAL

2.2.1 设计主题 40
DESIGN THEME

2.2.2 设计目标 43
DESIGN GOAL

2.3 设计策略 44
DESIGN STRATEGIES

3 规划设计及建成效果 46
PLANNING & DESIGN AND BUILT EFFECT

3.1 规划方案 48
PLANNING CONCEPT AND PLAN

3.1.1 设计概念及方案 48
DESIGN CONCEPT AND PLAN

3.1.2 功能分区规划 50
FUNCTIONAL ZONING

3.1.3 交通系统规划 52
TRANSPORT PLANNING

3.1.4 水系规划 62
WATER SYSTEM PLANNING
3.2 节点设计 64
NODES DESIGN
3.2.1 东入口 66
EAST ENTRANCE
3.2.2 北入口 70
NORTH ENTRANCE
3.2.3 南入口 86
SOUTH ENTRANCE
3.2.4 景点一 浮光揽月 100
FLOATING MOON LAKE
3.2.5 景点二 赵都新韵 112
THE NEW HANDAN PAVILION
3.2.6 景点三 山水邯郸 144
HANDAN PAVILION
3.2.7 景点四 青山画卷 160
FLOWERING TERRACES
3.2.8 景点五 工业遗风 166
INDUSTRIAL HERITAGE PAVILION
3.2.9 景点六 清渠如许 172
WETLAND TERRACES
3.2.10 景点七 矿坑花园 196
THE RECLAIMED QUARRY PARK
3.2.11 景点八 梦泽飞虹 217
THE FLYING BRIDGE
3.2.12 景点九 印塔夕照 228
HANDAN TOWER IN THE SUNSET
3.2.13 景点十 民俗印象 240
FOLK CULTURE PAVILION

4 建成感悟 267
REFLECTIONS
4.1 前期设计阶段 268
DESIGN STAGE
4.2 施工配合阶段工作心得 270
REFLECTIONS OF CONSTRUCTION STAGE
4.2.1 建设难点 270
CONSTRUCTION DIFFICULTIES
4.2.2 景观难点（以清渠如许景区为例） 272
DESIGN DIFFICULTIES (TAKING WETLAND TERRACES AS EXAMPLE)
4.2.3 建筑难点（以山水邯郸及涧沟陈展馆为例） 276
ARCHITECTURE DIFFICULTIES (TAKING HANDAN PAVILION AND JIANGOU MUSEUM AS EXAMPLE)

土人设计著作系列 279
PAST PUBLICATION OF TURENSCAPE DESIGN

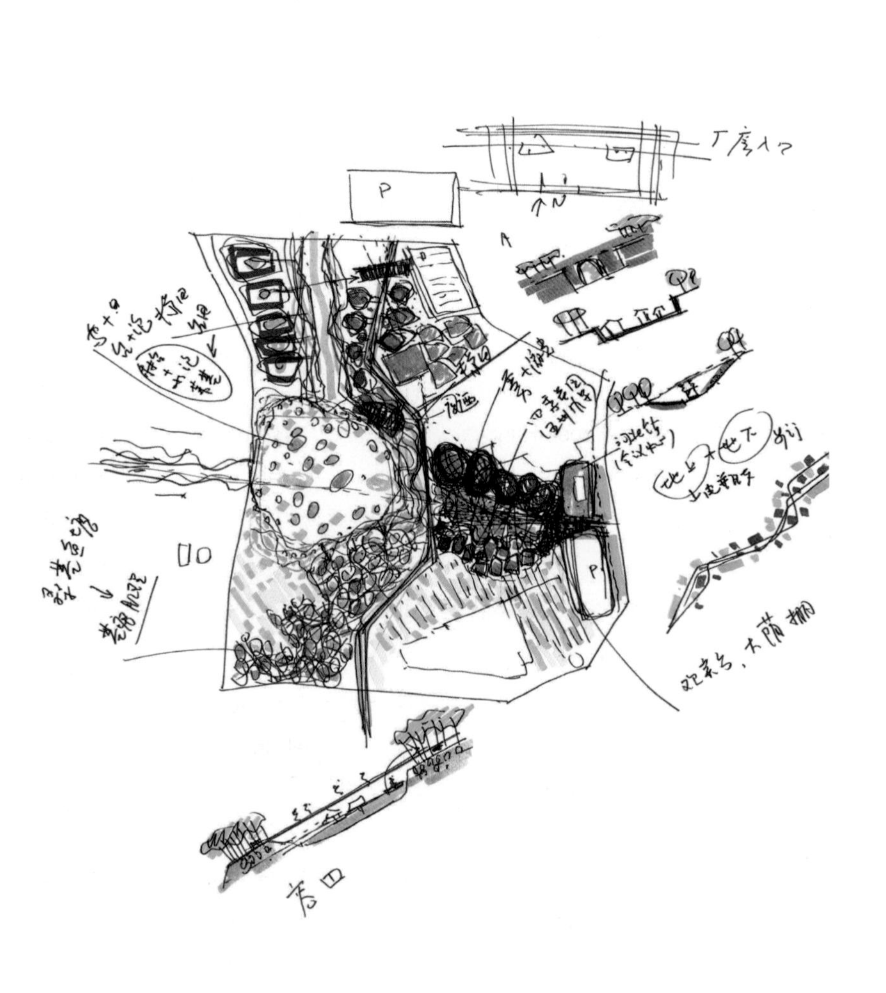

2018 年 11 月 30 日 首席设计师俞孔坚草图
Sketch drawing by chief designer Yu Kongjian, November 30 2018

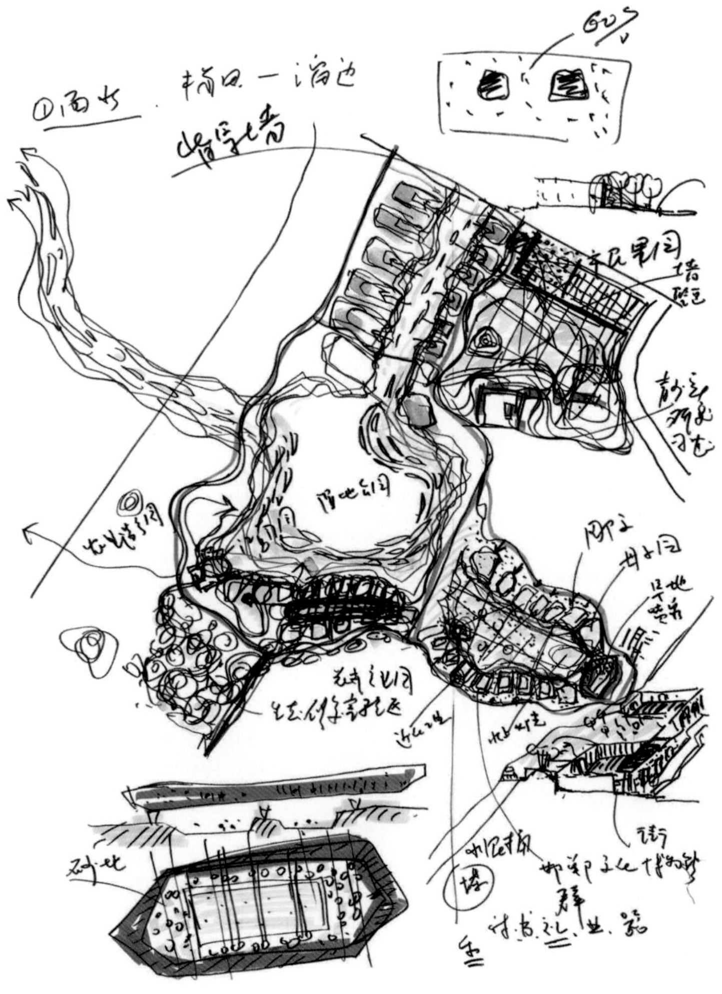

2019 年 3 月 10 日 首席设计师俞孔坚草图
Sketch drawing by chief designer Yu Kongjian, March 10 2019

概要
ESSENTIALS

作为推动城市创新发展、绿色发展、高质量发展的重要平台，河北省园林博览会已成功举办三届。河北省第四届（邯郸）园林博览会（以下简称邯郸园博会，会址简称邯郸园博园）由北京土人城市规划设计股份有限公司设计，从现场调研（2018年12月）到基本建成（2020年9月），历时近两年，前后经过了多次修改，让曾经4300亩（约287hm^2）邯钢工业棕地，蜕变为一处山水共融的绿色公园、城市生态修复的成功典范。

邯郸园博园位于邯郸市复兴区，是邯郸的母亲河——沁河汇入邯郸市的源头，后因城市扩张以及工业发展，这里变成垃圾遍地、臭水横流的工业废弃地。我们以“山水邯郸，绿色复兴”为设计主题，用当代的技术、当代的设计、当代的语言，尝试修复生态基底，重塑绿色环境，践行“海绵城市”与“城市双修”的理论，让复兴区实现由“工业污染区”向“绿色生态区”的华丽转身，大幅度提升所在区域的蓝绿空间占比。不仅满足河北省第四届园林博览会的使用需求，还补充了邯郸城市缺少大规模公共绿地的功能短板，优化了城市空间布局，成为生态秀美、业态丰富的城市新发展引擎。在“园博效应”的带动下，以“园”带“城”，为复兴区和邯郸市的商贸、旅游发展带来新的机遇。

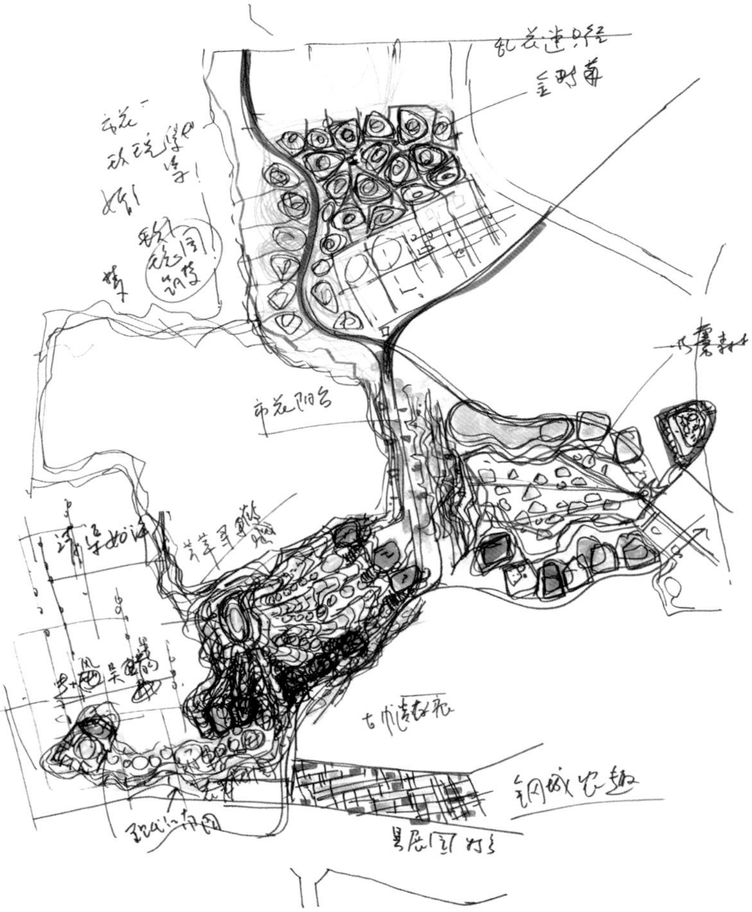

As an important platform for cities' innovative development, green development and high-quality development, the Hebei Garden Expo has been successfully held for three sessions. The Fourth Hebei (Handan) Garden Expo (hereinafter: Handan Garden Expo and its venue as: Handan Garden Expo Park) was designed by Beijing Turenscape Urban Planning and Design Co., Ltd. It took nearly two years from on-site investigation (December 2018) to completion (September 2020), undergone many revisions, to turn the 4300 mu (287 hectares) of the industrial brownfield of Handan Steel into a green park, setting a successful example of urban ecological restoration.

Handan Garden Expo Park is located in Fuxing District, Handan City, where Qin River flows into it. Due to urban expansion and industrial development, it became an industrial wasteland with rubbish and stinking water. We took “Natural Handan, Green Revival” as design theme, using contemporary design techniques, and language to restore the ecological base, reshape the green environment, and practice the theories of “sponge city” and “city betterment, ecological restoration”. The design allowed the district a gorgeous turnaround from an “industrial pollution area” to a “green ecological area”, greatly increased the blue and green space in the area. It surpassed the use needs of the Fourth Garden Expo in Hebei Province, and increased large public green space in the city, optimizes the urban layout, making the park become a new development engine with beautiful ecology and rich business formats. Driven by the “Garden Expo Effect”, the “Garden” will lead the “City” to bring new opportunities for the development of commerce and tourism in both Fuxing District and Handan City.

2019年3月20日 首席设计师俞孔坚草图
Sketch drawing by chief designer Yu Kongjian, March 20 2019

1 项目概况
PROJECT OVERVIEW

1.1 项目背景

PROJECT BACKGROUND

近代以来，得益于优越的资源禀赋和良好的区位条件，邯郸成为中国近代工业的主要发祥地之一，被誉为“钢城”和“煤都”。随着资源的过度开采，传统产业萎缩，早期城市扩张和工业发展带来的后遗症逐步显现：老城区生态空间匮乏、市政基础设施薄弱，特别是空气污染问题，成为困扰邯郸城市发展的一个顽疾，城市与自然、人与水之间的矛盾日益凸显。

党中央提出生态文明理念，要求正确处理经济发展同生态环境保护的关系，牢固树立和践行“绿水青山就是金山银山”的理念，满足“人民对美好生活的向往”，加强人与自然和谐关系纽带的建立。

After 1910s, thanks to its superior resource endowment and locational conditions, Handan has become one of the main birthplaces of the modern China's industry, known as the “city of steel” and the “city of coal”. With the sequent over-exploitation of resources and the shrinking of traditional industries, the sequelae caused by early urban expansion and industrial development have gradually emerged: lack of ecological space in the old city, weak municipal infrastructure, air pollution etc. have become stubborn diseases that plague the urban development of Handan. The contradictions between city and nature, people and water have become increasingly prominent.

The Party Central Committee put forward the concept of ecological civilization, requiring correct handling of the relationship between economic development and ecological environmental protection, firmly establishing and practicing the concept of “Clear waters and green mountains are as good as mountains of gold and silver”, satisfying “people's yearning for a better life”, and strengthening a harmonious relationship between people and nature.

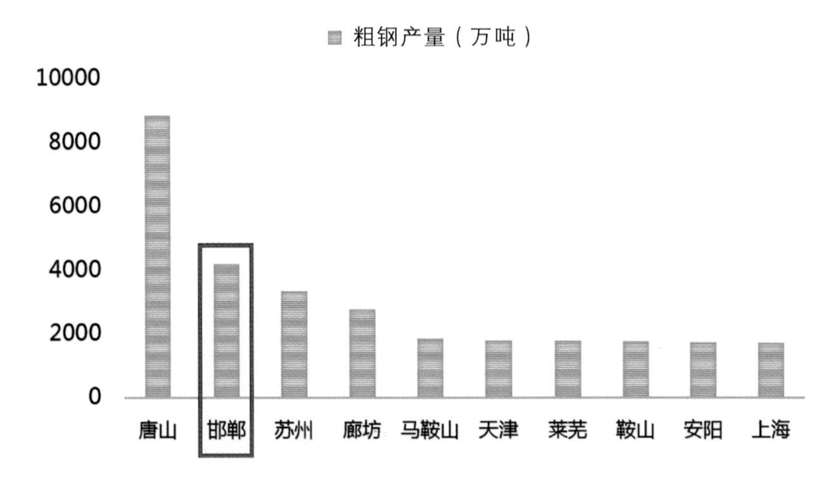

图 1-001 2016 年全国各城市（粗钢）产量前十城市排名
Figure 1-001 National ranking of the top 10 crude steel production by cities in 2016

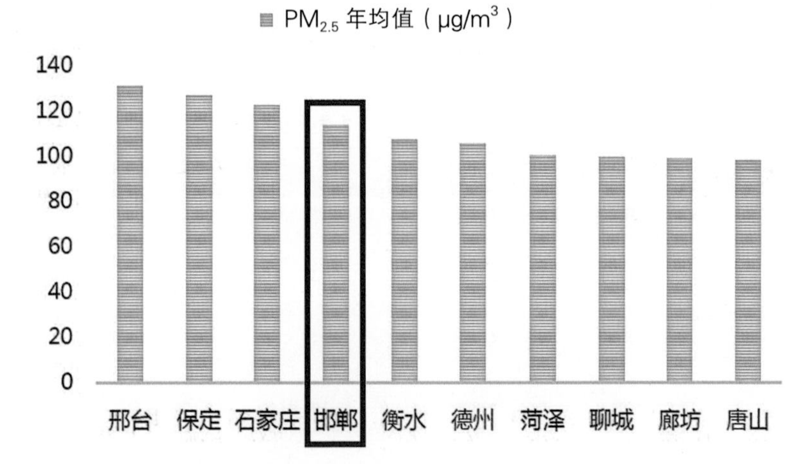

图 1-002 2017 年全国雾霾前十城市排名
Figure 1-002 Top 10 most hazardous city in China in 2017

图 1-003 邯郸市城市扩张发展趋势
Figure 1-003 Urban sprawl and development of Handan

图 1-004 邯郸市工业斑块及绿色斑块分布图
Figure 1-004 Distribution of industrial and green patches in Handan

图 1-005 邯郸市空气污染实景
Figure 1-005 Photos showing air pollution in Handan

河北省第四届（邯郸）园林博览会选址于邯郸市近郊的重、化工业区，并临近南水北调工程，处于西侧山水生态环境及复兴重工业区交界的关键位置，生态地位重要。环城水系以及绿廊将串联园博园以及北、东、南三个主要绿地，形成邯郸市城市绿环，利于改善城市环境，促进城市结构转型与城市复兴。

The Fourth Hebei (Handan) Garden Expo is located in the heavy and chemical industrial zone in the suburbs of Handan City, which is close to the South-to-North Water Diversion Project. It is located at the junction of the landscape ecological zone on the west and the Fuxing heavy industrial zone, featuring significant ecological status. The water system around the city and the green corridor connect the Garden Expo Park with the three major green spaces in the north, east and south, forming a green ring for Handan which improves the urban environment, and promotes urban structural transformation and urban regeneration.

图 1-006 由园博园看向东部城市风貌
Figure 1-006 VIewing the city to the east from the garden Expo

图 1-007 邯郸市山水生态格局
Figure 1-007 Ecological layout of Handan city

1.2 文化脉络

CULTURAL TRADITIONS

邯郸自上古时期至今历史源远流长，涵盖多个朝代，历经 8000 多年的文化积淀，形成以“赵文化”为核心的女娲文化、磁山文化、邺城文化、石窟文化，以及广府太极文化、梦文化、磁州窑文化、成语典故文化和红色文化等十大文化脉络，是园博园设计工作的主要文化线索。

Handan has a long history since ancient times and undergone multiple dynasties. After more than 8,000 years of cultural reform, Handan has formed Nuwa Culture, Cishan Culture, Yecheng Culture, Grotto Culture, and Guangfu Tai Chi Culture, with “Zhao Culture” as the core, and dream culture,Cizhou kiln culture, idiom allusion culture and red culture. These ten cultural traditions are the main cultural clues of the expo design.

远古时期	史前时期	夏商周时期	春秋战国时期	汉朝	三国两晋南北朝	隋唐五代时期	宋辽金元时期	明清时期
传说上古时期人类始祖女娲就在邯郸古中皇山抟土造人、炼石补天，邯郸是**中国早期文明发源地。**	邯郸为**磁山文化**发祥地，武安磁山是我国发现的一处新的新石器时代早期文化遗址，距今约7300年，是**中国早期文明发源地。**	**邯郸城邑**最早出现于商末的纣王时期，距今至少已有3100年。今邯郸的漳水流域便是远古时期商族的发祥之地，考古界称之为**“下七垣文化”**，也称先商文化。	春秋前期，邯郸先属卫后属晋，春秋后期被晋国赵氏宗主夺取。公元前386年，**赵国迁都邯郸**，邯郸作为赵都历经8世，达158年之久。	西汉时期飞速发展，西汉后期发展为**五都之一**，邯郸之战后逐渐衰落，隶属魏郡。	三国属广平郡，东晋后改属魏郡。	先后归属或复辖为洺州、磁州、武安郡和紫州，衰落而成蕞尔小县。	宋金时属洺州，元代县属河北路磁州，大名为河北路治所。元代**水利有所发展。**	属北直隶省 平府，政治 济中心在广 府城。明代 后水运有所 展。

图 1-008 邯郸历史沿革
Figure 1-008 Urban transformation of Handan

国时期　　**中华人民共和国**

国初直隶
冀南道，
8年，直
改为河
。

1954年改为省辖市。1972年**磁山文化**被发现，从而揭开了黄河流域早期新石器文化探索的序幕。

图 1-009 邯郸十大文化脉络
Figure 1-009 10 dominant cultural traditions in Handan

1.3 外部交通

TRANSPORT ACCESSIBILITY

邯郸园博园周边交通便利，区位优势明显。场地东侧紧邻西外环路，北临 G309 新邯武公路，南至邯武快速路，离机场、邯郸站等重要交通节点的距离适中。

邯郸市游客群主要出行工具为公共交通及自驾。外省市游客主要通过高铁站点、自驾方式到达邯郸市区，再通过西环路到达园博园。

With its advantageous location, Handan Garden Expo Park boasts convenient traffic accessibility. The site adjoins the West Outer Ring Road on the east side, the new G309 Handan–Wu'an Highway on the north side, and Handan–Wu'an Expressway on the south side, with a moderate distance from important transportation nodes such as the airport and Handan Railway Station.

The main means of travel for tourists in Handan City are public transportation and driving. Tourists from other provinces arrive in Handan mainly through the high-speed train station and by driving, and then take the West Ring Road to arrive at the Park.

图 1-010 邯郸园博园周边交通关系图
Figure 1-010 Transportation system surrounding the Handan Expo campus

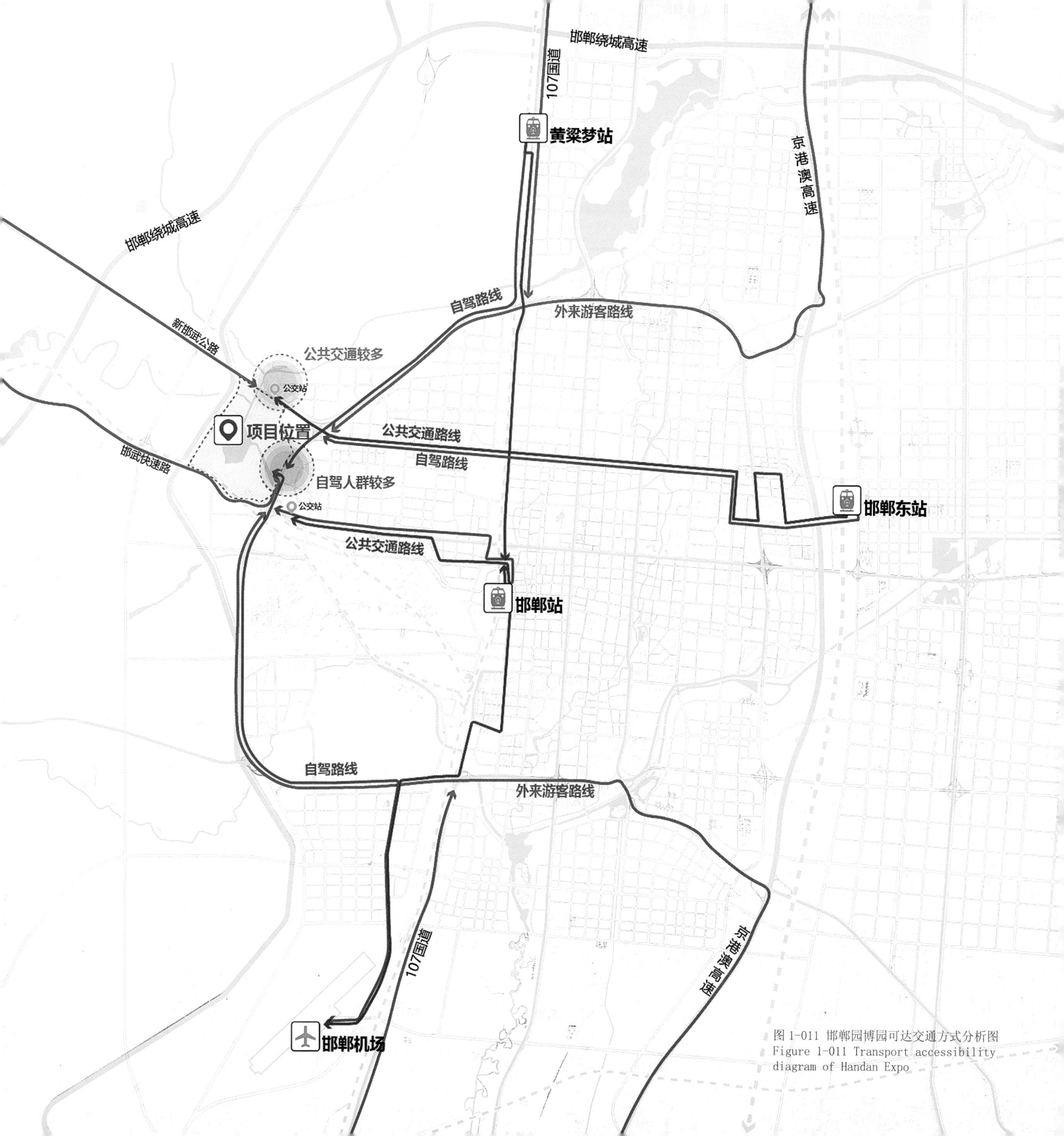

图 1-011 邯郸园博园可达交通方式分析图
Figure 1-011 Transport accessibility diagram of Handan Expo

1.4 场地认知

Site Cognition

邯郸园博园总规划面积约 4242 亩（约 283hm²），核心游览区面积约 1838 亩（约 123hm²）。场地内地势高低起伏，竖向变化大，同时，还有西湖水库、涧沟遗址等自然文化遗存。此外，场地内大量的邯钢工业废弃物，对土地、河水有较大污染，环境问题突出。

The total area of the project is about 4242 mu (283 hectares), and the core tourism area is about 1838 mu (123 hm²). The terrain in the site is undulating, with great elevation change. There are natural and cultural relics within the area including the Xihu Dam and the Jiangou Ruins. In addition, Handan steel's industrial wastes in the site polluted the land and river, causing serious environmental problems.

图 1-012 场地内整体景观特征及其分布
Figure 1-012 Overall landscape features and their distributions on site

1.4.1 自然条件
NATURAL CONDITIONS

（1）水文条件
HYDROLOGIC CONDITION

场地内的西湖水库是环城水系规划的四大湖区之一，连接沁河流域与北湖、南湖、东湖以及多条河流，是环城绿环的重要节点，城市西部的生态屏障，也是邯郸的母亲河——沁河汇入邯郸市的首要门户，在生态、防洪、净化等多方面起到重要作用。

The Xihu Dam in the site is one of the four major lake areas planned for the water system around the city. It connects the Qin River Basin with the North Lake, the South Lake, the East Lake and many rivers. It is an important node in the green ring around the city, the ecological barrier in the west of the city, and the main gateway for Handan's Mother River——the Qin River. It plays an important role in many aspects such as ecology, flood control, and purification.

图 1-013 西湖水库现状实景
Figure 1-013 Photo showing existing condition of the Xihu Dam

图 1-014 邯郸市水系规划图
Figure 1-014 Hydrological planning diagram of Handan City

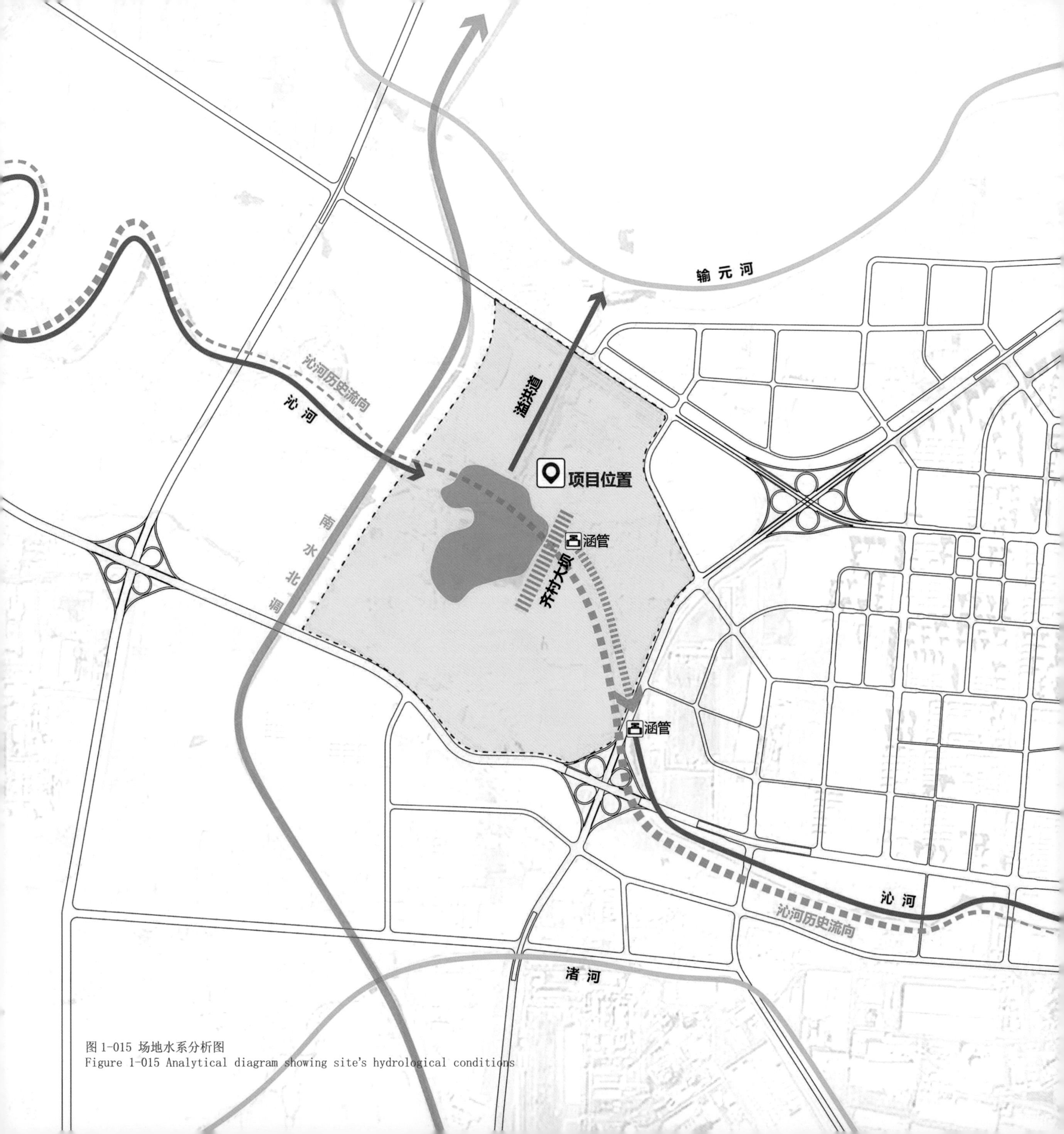

图 1-015 场地水系分析图
Figure 1-015 Analytical diagram showing site's hydrological conditions

沁河流入场地内，齐村大坝的分隔及设立的溢洪道，使得沁河水系整体流向输元河，仅有少量水流通过涵管流向沁河下游，部分继而流入城区。

The Qin River flows into the site. With the Qicun Dam and the affiliated spillway, the majority of the Qin River system flows to the Shuyuan River. Only a small amount of water flows through the culvert to the lower reaches of the Qin River, part of which flows into the city.

齐村大坝西侧为西湖水库，湿地滩涂态势良好，但是整体水质需要改善。东侧原河道成为邯钢水渣场矿渣堆砌地，阻隔了地表水系连通，污染严重，亟待在本次建设中予以整治。

On the west side of the Qicun Dam is the Xihu Dam. Although the wetland tidal flats are in good condition, the overall water quality is in need of improvement. The original river channel on the east side had become the slag piling ground of Handan Steel's water slag field, which blocked the connection of the surface water system. The serious pollution problems in the site urgently needed to be rectified by the project.

图 1-016 齐村大坝及水利塔现状实景
Figure 1-016 Photo showing existing condition of the Qicun Dam and the water tower

（2）地形地貌

TOPOGRAPHY

场地地形丰富多变，整体呈现西北高、东南低的态势，海拔 78 ~ 116m。齐村大坝拦截沁河河道，两侧地形在原河道范围内高程较低，形成较低地势。场地西南区域散布历史墓地遗迹，形成突出地表的小坡地。

Boasting rich and changeable terrain, the overall site is higher in the northwest and lower in the southeast, with the altitude ranging between 78m and 116m. The Qicun Dam intercepts the Qin River channel, and the terrain on both the banks lie on a rather low elevation within the original channel. In the southwest area of the site, relics of historical cemeteries are scattered to form small bumps on the ground.

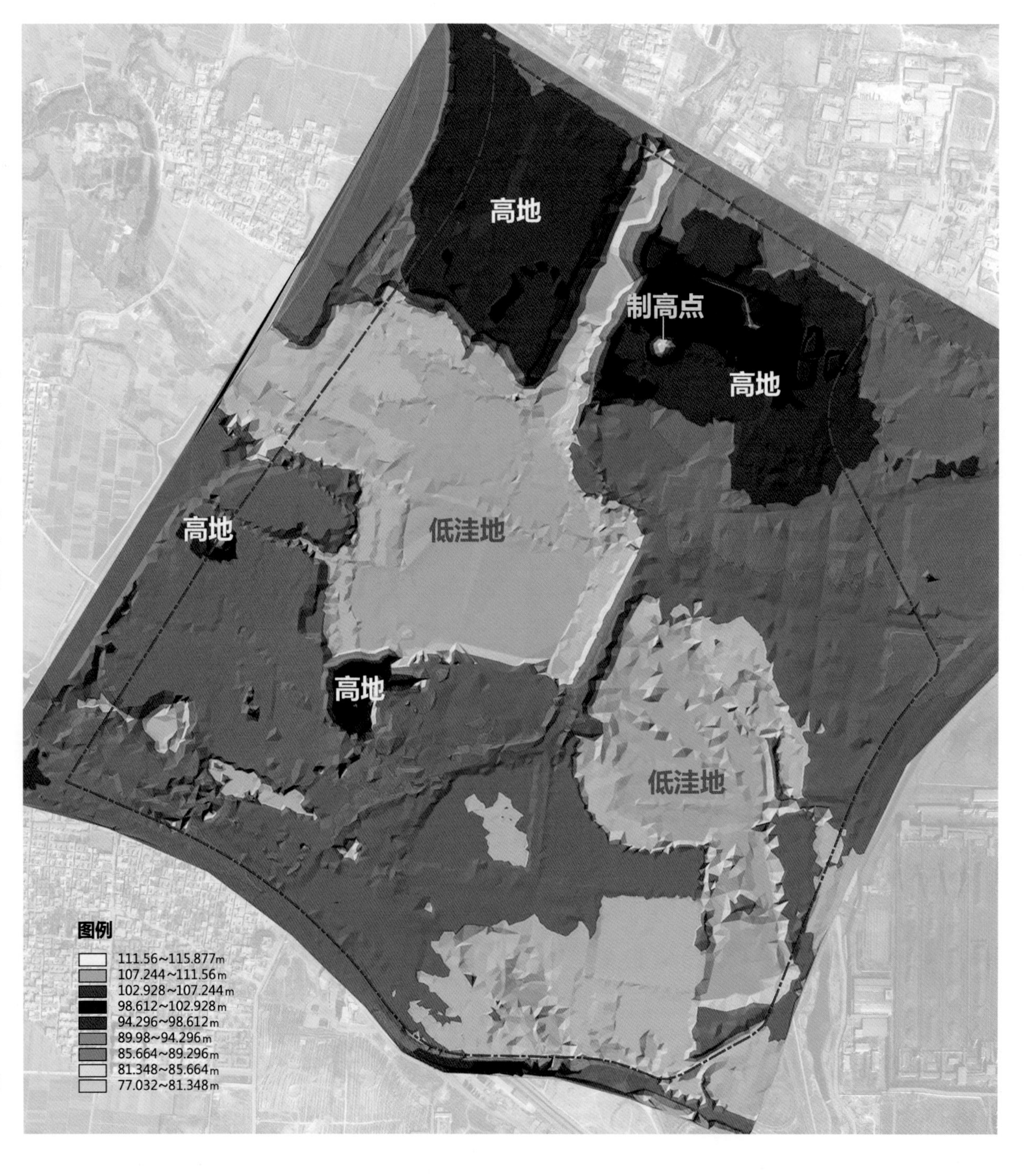

图 1-017 场地地形地貌分析图
Figure 1-017 Analytical diagram of site typography

（3）现状植物

EXISTING VEGETATION

作为城市工业废弃地，场地内多为闲置空地，植被资源分布不均。北部片区多为林地和园地，其中沿溢洪道坡地上有大量现状树木，但树木形态与生长情况良莠不齐。西侧库区范围内为大片现状湿地，水生植被条件较好，但南侧驳岸受污染影响，植被有所退化。场地其他区域范围内棕地污染相对严重，植被条件较差，多为硬质水泥场地及裸露的土壤。

As an urban industrial wasteland, the site is mostly idle with unevenly distributed vegetation resources. The northern area is mostly covered by forest land and garden land. There is a large number of existing trees on the slopes along the spillway, but the morphology and growth condition of the trees are uneven. The reservoir area in the west is an existing large wetland with good aquatic vegetation conditions, however its southern revetment is affected by pollution, and thus the vegetation has been degraded. In other areas of the site, brownfield pollution seemed serious, with poor vegetation conditions, mostly hard cement ground and bare soil.

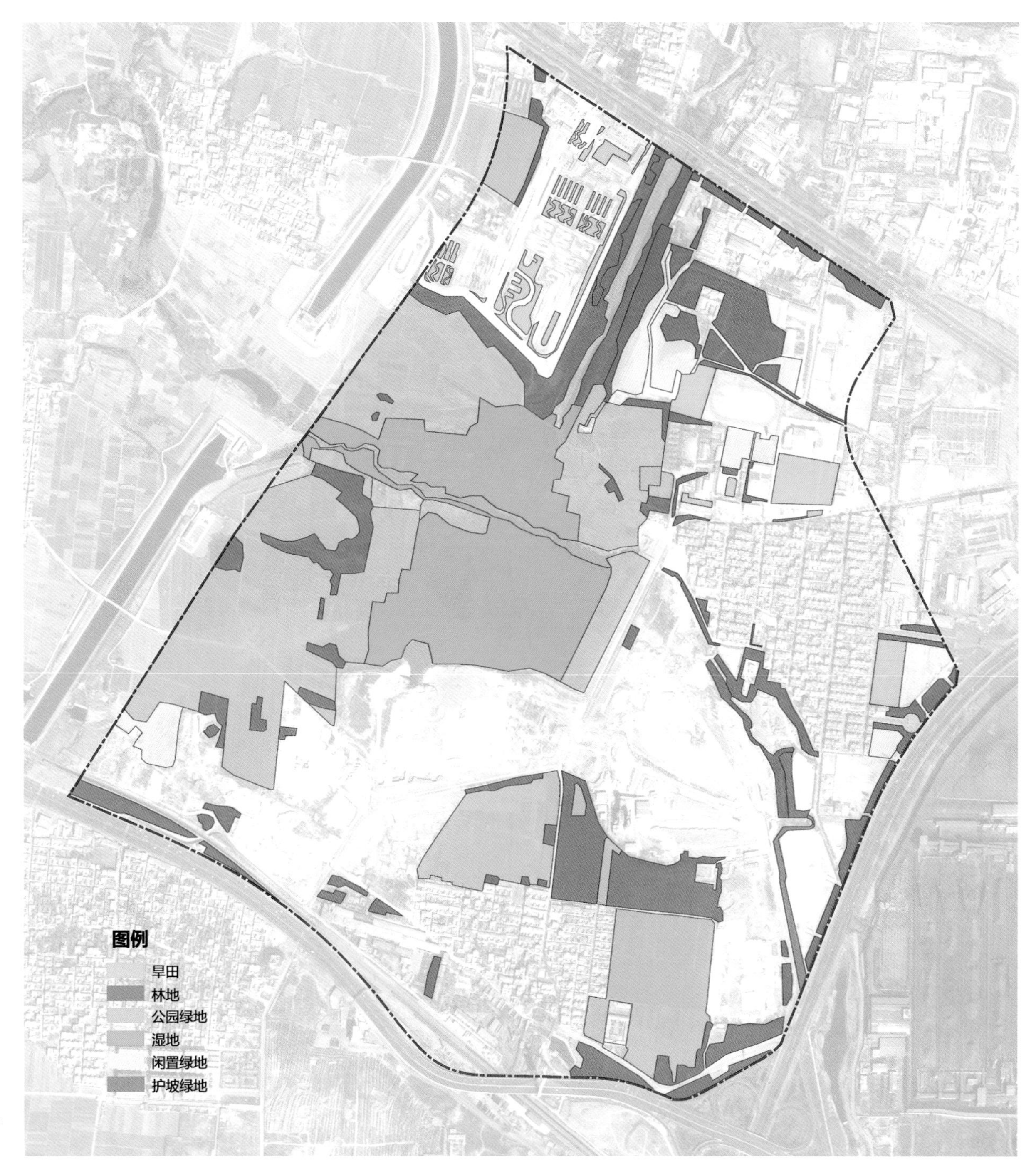

图 1-018 场地现状植物分析图
Figure 1-018 Analytical diagram of existing plantations onsite

1.4.2 人文资源

Cultural Resources

场地南翼为洞沟古村，村北为涧沟遗址，是邯郸城的最早雏形，有房址、水井、陶窑和丛葬坑等文化遗迹，被国务院公布为第七批全国重点文物保护单位，对于研究龙山文化有重要意义，是场地不可忽视的重要文化资源。其中涧沟遗址水井是中原地区迄今发现的年代最早、结构最复杂的水井，为研究中国水井和“井”字的起源提供了十分宝贵的资料。

To the north side of the site lies the Jiangou ancient village and the north of which is the Jiangou ruins, the earliest prototype of Handan City. There are cultural relics such as house sites, water wells, pottery kilns and Cong burial pits. Listed by the State Council within the seventh batch of National Key Cultural Relics Protection Units. It is of great significance to the study of Longshan culture, and therefore an important cultural resource that cannot be ignored. Among them, the Jiangou ancient water well is the earliest and most complex water well discovered in the Central Plains so far, which provides valuable data for the study of the origin of Chinese water wells and the word “井” (Well).

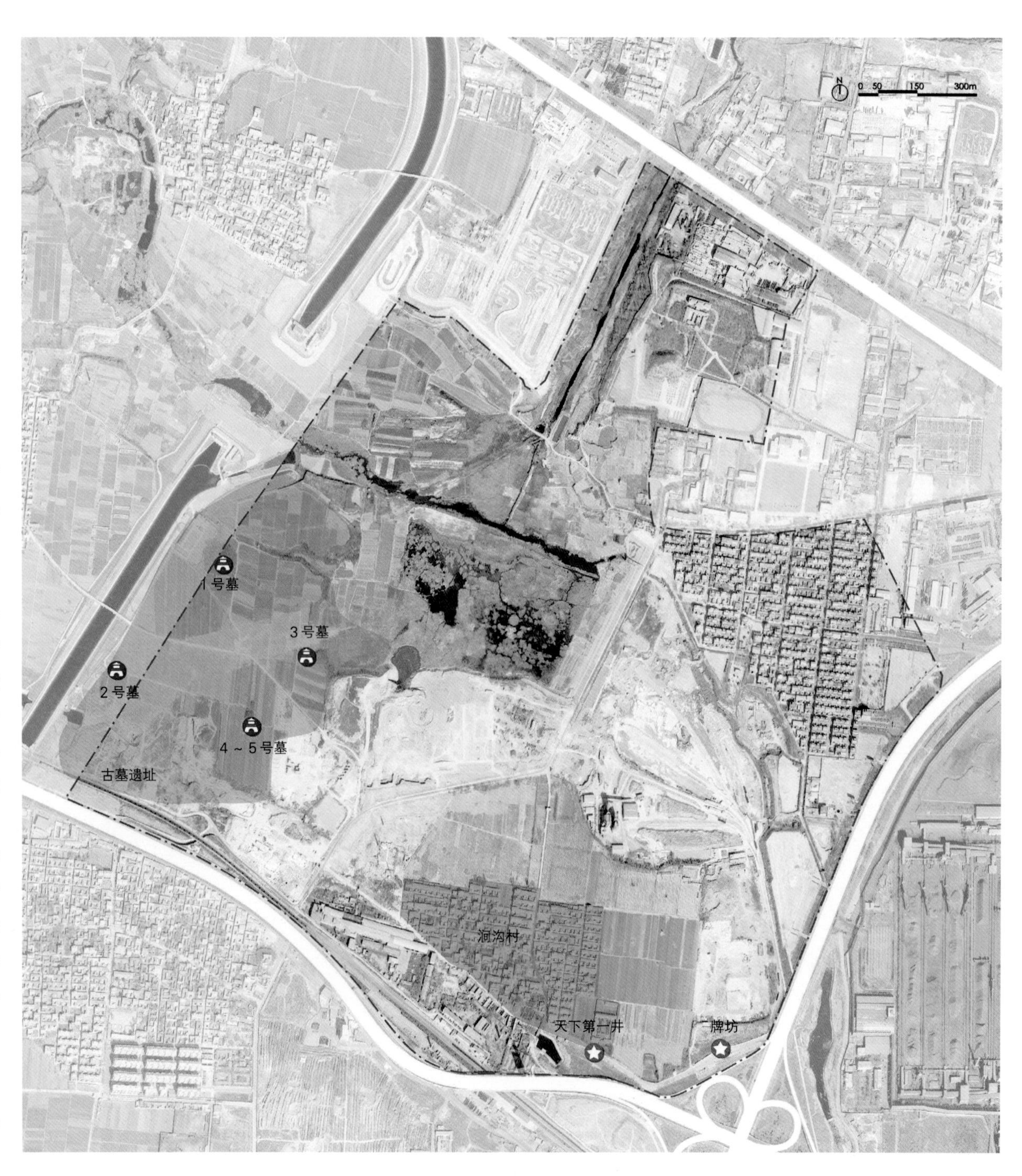

图 1-019 场地人文资源分布图
Figure 1-019 Human resources distribution map

场地西侧，紧邻南水北调处为现状林村墓群，是战国、汉代的古墓葬遗址，是邯郸市复兴区的一个全国重点文物保护单位。

On the west side, adjacent to the South-to-North Water Diversion Project, locates the existing Lincun cemeteries. Being an ancient tomb site of the Warring States and the Han Dynasties, it is a national key cultural relic protection unit in the Fuxing District.

图 1-020 涧沟村现状实景
Figure 1-020 Current photos of Jiangou village

图 1-021 涧沟遗址——水井现状实景
Figure 1-021 Jiangou ruination-existing condition photo of water well

图 1-022 林村墓群遗址现状实景
Figure1-022 Existing condition photo of Lincun Cemeteries

1.4.3 限制因素
CONSTRAINT CONDITIONS

场地限制因素包括南水北调工程、齐村大坝、水利设施、涧沟村文物遗址、墓葬遗址、基本农田等。

The site constraints include the South-to-North Water Diversion Project, the Qicun Dam, water conservancy facilities, the Jiangou Village cultural relics, tomb sites, and prime farmland.

图 1-023 场地限制因素分析图
Figure 1-023 Analytical diagram of site constrains

N
0 50 150 300m
天线电塔
天线电塔
军械修理所
齐村大坝
燃气管线
齐村污水站
建设控制地带
涧沟遗址
涧沟污水站
军械修理所
齐村大坝
全国重点文物保护单位
涧沟遗址
中华人民共和国国务院二零一三年五月三日 公布
河北省人民政府二零一五年七月十日 立

2 挑战与策略

CHALLENGES AND STRATEGIES

2.1 建设挑战

CONSTRUCTION CHALLENGES

邯郸园博园场地虽资源丰富，但是生态环境十分脆弱，土壤水质污染和废弃工厂遗留工业垃圾等环境问题十分严重，而且复杂的竖向变化在给予设计师充分发挥空间的同时，也使项目更具复杂性及综合性，需要与专业人员共同解决面临的问题，这些都是对设计师的巨大挑战。

Although the site is rich in resources, its ecological environment is very fragile. Environmental problems such as soil and water pollution and industrial waste from abandoned factories are serious. Moreover, while the complex elevation change gives designers more space for creation, it makes it more complex and requires coordination with various professionals to solve the problems faced, bringing huge challenges for the designers.

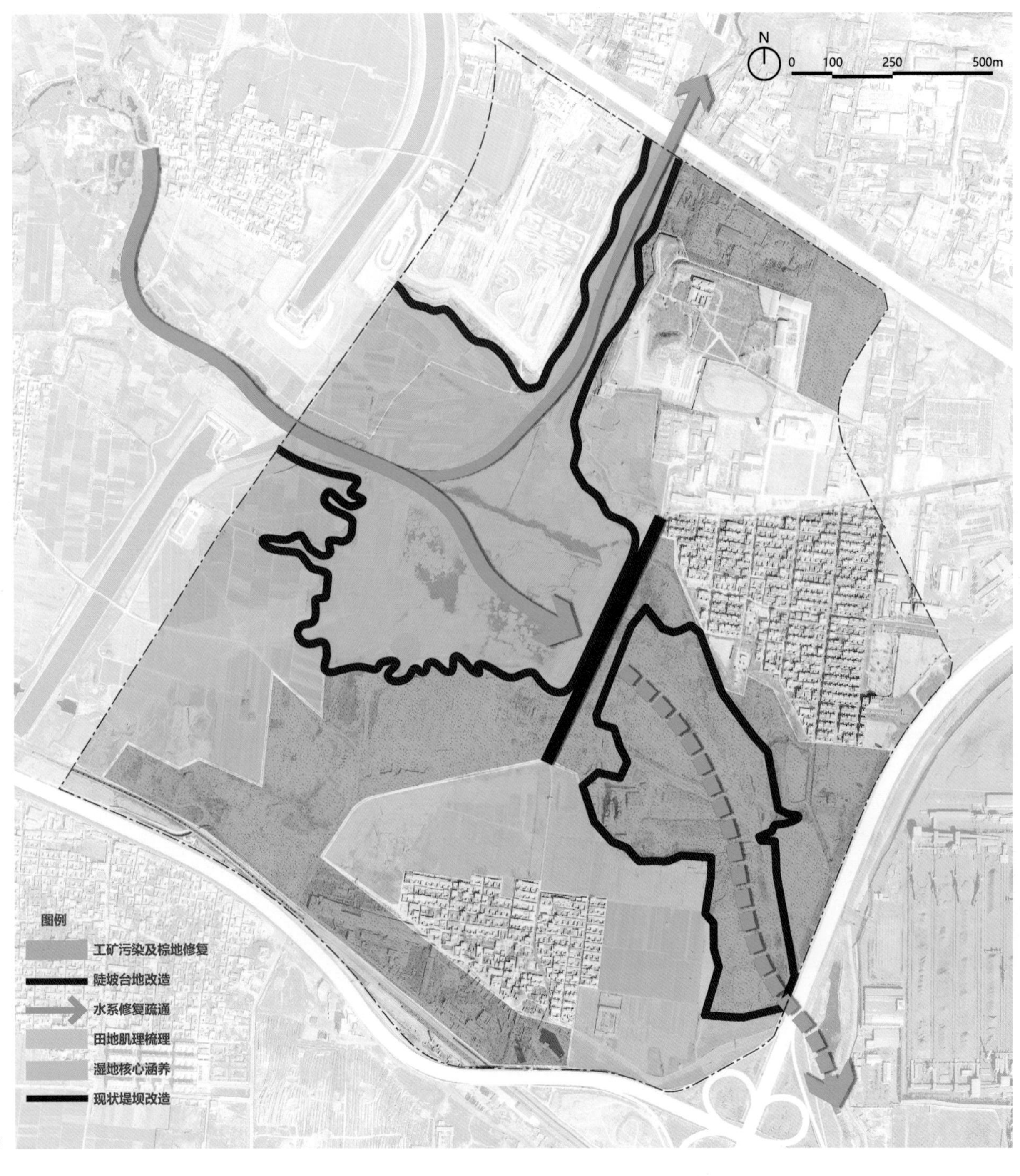

图 2-001 生态修复类型分布图
Figure 2-001 Distribution of different ecological restoration types

场地齐村大坝东侧为水渣场废弃地，除了大量工业废弃物、作业设备、机器等，还遗留有大面积的含有超量的铬、氮和磷的超厚层废弃矿渣，严重污染周边水体，同时土地硬化明显，这些都需要在后续工作中提出系统处理方案。

A waste slag field lies on the east side of Qicun Dam where, in addition to a large amount of industrial waste, operating equipment, machinery, etc., a large area of waste slag containing excessive amounts of chromium, nitrogen, and phosphorus seriously polluted the surrounding water bodies. The land became obviously hardened, requiring a systematic treatment plan.

图 2-002 齐村大坝东侧现状生态问题
Figure 2-002 Current ecological issues of north Qicun Dam

齐村大坝西侧为库区，水质较差，水体富营养化情况严重。

The reservoir on the west of Qicun Dam is in poor water quality and serious eutrophication.

图 2-003 西湖水库现状生态问题
Figure 2-003 Current ecological issues of Xihu Dam

库区南侧为工业矿渣和建筑垃圾堆叠而成的“小山”，是全园环境问题最为严重的区域，靠近邯武快速路的区域还有一处人工采石场遗迹，形成一个巨大的矿坑。

On the south side of the reservoir area, there is a “hill” made up of industrial slag and construction waste. It is the area with the most serious environmental problems in the whole park. In addition, there is a relic of an artificial quarry near the Handan–Wu’an Expressway, which forms a huge pit.

图 2-004 西湖水库南侧现状生态问题
Figure 2-004 Current ecological issues of north Xihu Dam

2.2 设计主题与目标

DESIGN THEME AND GOAL

2.2.1 设计主题

DESIGN THEME

本届园博会的主题是“山水邯郸·绿色复兴”，传承往届创新精神，倡导生态文明新风。

山水邯郸：“壮丽太行山川，碧透滏阳河水，千年古迹韵味浓厚，秀美村庄风景如画。”邯郸不仅文化底蕴深厚，而且山川秀美、景色宜人。邯郸园博园重塑邯郸的“山、水”重要城市特色要素，让山水、古城相融相连，推动邯郸区域转型发展。

绿色复兴：“复兴”既点明本届园博会的承办地，又将毛主席于 1959 年 9 月视察邯郸时指出“邯郸是要复兴的”与当代邯郸城市转型需求融合，将生态文明建设确定为邯郸再复兴的一个标志。从工业文明走向生态文明，从一个垃圾遍地、臭水横流、臭气熏天的工业废弃地变成承载人民美好生活向往的美丽花园，不仅是复兴区的复兴、邯郸的复兴、河北的复兴，更是全国城市复兴的典范。

The theme of this year's garden expo reads “Natural Handan, Green Revival”, inheriting the innovative spirit of the past and advocating a new style of ecological civilization.

Natural Handan: “The magnificent Taihang mountains and rivers, the clearest water of Fuyang River, the thousand-year-old monuments with lasting charm, and beautiful villages lie in picturesque landscape.” Handan boasts not only profound cultural heritage, but also beautiful scenery. The Handan Garden Expo Park reshapes important urban characteristics of Handan's landscape, integrates landscape and the ancient city, and promotes the regional transformation and development of Handan.

Green Revival: “Fuxing” (the word “Fuxing” in Chinese means “revival”) not only points out the revenue of the expo, it also integrates Chairman Mao's statement “Handan is going to be rejuvenated”, when he inspected Handan in September 1959, with the needs of the contemporary Handan's urban transformation, determining ecological civilization as a sign of Handan's rejuvenation. From industrial civilization to ecological civilization, from an industrial wasteland filled with rubbish and stinky water to the beautiful garden for which people yearn. It is not only the revival of Fuxing District, Handan City, or Hebei Province, it is a model of the national urban rejuvenation.

图 2-005 鸟瞰图（主入口视角）
Figure 2-005 Bird eye （viewing from the main entrance）

图 2-006 鸟瞰图（南入口视角）
Figure 2-006 Bird eye （viewing from the north entrance）

2.2.2 设计目标

Design Goal

邯郸园博会的设计目标是“重建人与自然的和谐共生，打造永不落幕的园博会”。

实现“人民对美好生活的向往”：将园博会的建设与市民生活的需求结合在一起，体现中央执政为民的理念。

打造“山水林田湖草生命共同体”：对场地内遭到不同程度破坏的生态环境进行全方位、彻底的修复，将沁河、园博园与城市连为一体，打造山水林田湖草生命共同体。

践行“绿水青山就是金山银山”：让园博园的建设，成为邯郸城市转型、乡村转型和文化振兴的主要抓手，实现经济增长、土地增值的城市发展目标。

The design goal of Handan Garden Expo is to “rebuild the harmonious coexistence between man and nature, and create a garden expo that never ends”.

To realize “the people’s yearning for a better life” : Combine the construction of the park with the citizens’ needs, and reflect the central government’s concept of “Governing for the People”.

To create “a living community of landscape, forest, field, lake and grass”: Restore the ecological environment in the site that has been damaged, connect the Qin River, the park and the city, and create a living community of landscape, forest, field, lake and grass.

To practice “Clear waters and green mountains are as good as mountains of gold and silver” : Make the construction of the park become the main focus of Handan’s urban–rural transformation and cultural revitalization, and realize the urban development goal of economic growth and land appreciation.

2.3 设计策略

DESIGN STRATEGIES

生态涵养，湿地保护

ECOLOGICAL CONSERVATION, WETLAND PROTECTION

现状西湖区域采用最小干预的生态设计理念，保留原有湿地肌理，在湿地边缘区域根据不同的场地情况进行适当的植被补植及景观提升，让改造后的湿地成为鸟类及多种生物的栖息场所。

In the existing Xihu area, a design concept of minimal intervention is adopted. The original wetland texture is preserved, with appropriate vegetation replanting and landscape upgrading carried out at the edge of the wetland according to different site conditions, transforming the wetland into habitat for bird and other species.

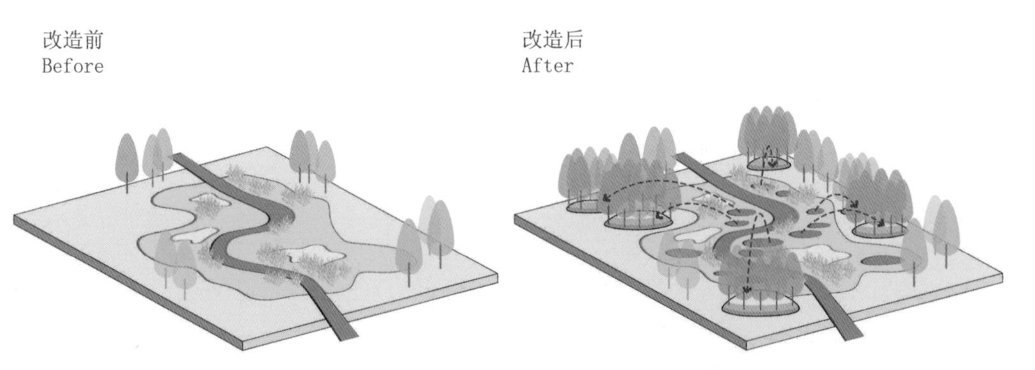

图 2-007 生态涵养、湿地保护策略模式图
Figure 2-007 Ecological conservation and wetland preservation model diagram

换填引水，土壤净化

LANDFILL AND WATER DIVERSION, SOIL PURIFICATION

设计采用三种不同的修复方式综合解决场地西部片区的土壤问题，包括矿渣填埋、土壤置换和引水净化。将现状矿渣进行填埋封存，对部分土壤进行置换，以满足园博园活动场地需求和植被生长的条件。在其他区域，引入湿地水系并进行自然水系生态净化。

The design adopts three restoration methods to solve the soil problems in the western part, including slag landfill, soil replacement and water diversion and purification. The existing slag was buried and sealed, Some soil was replaced to meet the needs of the expo and vegetation growth. In other areas, wetland water systems were introduced to enable ecological purification of natural water systems.

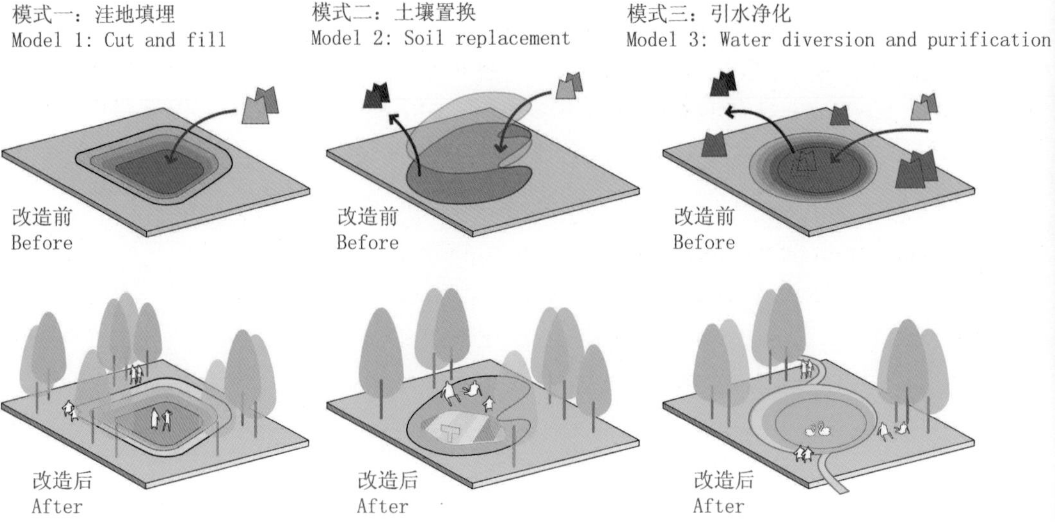

图 2-008 换填引水、土壤净化策略模式图
Figure 2-008 Slag landfill, soil remediation and water channeling model diagram

水系疏通，海绵修复

WATER SYSTEM DREDGING, SPONGE RESTORATION

针对齐村大坝以东区域，梳理并连通原有的沁河水道，利用现状矿坑洼地打造开阔的水面空间，湖水重新连接沁河的上下游，实现了邯郸的母亲河——沁河的系统性修复，从而形成核心游览区的中央湖体景观。沿园路布置雨水收集带和雨水花园，结合透水材料的运用，提高雨水的下渗和滞留，与湖体一同形成园博园的海绵体系。

For the area east to Qicun Dam, we reorganized the original Qin River waterway. Used the existing quarry pits to create open water surface. Thus restored the river through connecting the tributaries. And transformed the lake into a core tourism area. Along the paths, there are rainwater collection belts and gardens. The use of permeable materials increase infiltration and rainwater retention. Together with lake system, they form the sponge system.

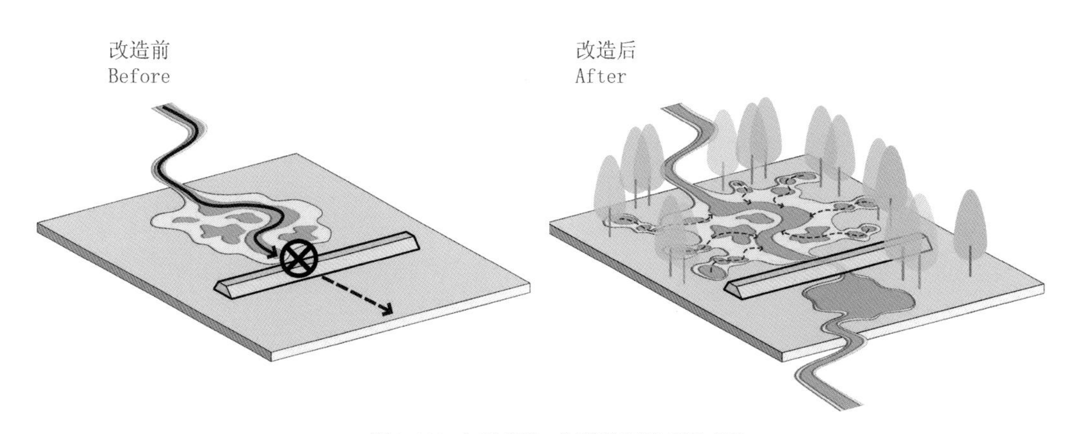

图 2-009 水系疏通、海绵修复策略模式图
Figure 2-009 Dredging and sponge restoration model diagram

陡坡填土，营造台田

STEEP SLOPES ENBANKMENT, TERRACED FIELDS

针对园区内齐村大坝以及西部矿坑陡坡区域，将原有陡坡层叠填土形成台田肌理，打造地形景观，并在台田上布置栈道、观景平台和景观盒，在利用原有地形高差的同时，形成宜游宜赏的台地景观效果，丰富游客的游园体验。

At the Qicun Dam and the quarry pits in the west side of the park, the original steep slopes are stacked and filled to form a terraced texture, and boardwalks, viewing platforms and landscape boxes are arranged on the terraces to make full use of the original height difference, forming a terraced landscape for touring and appreciation, enriching visitors' experience.

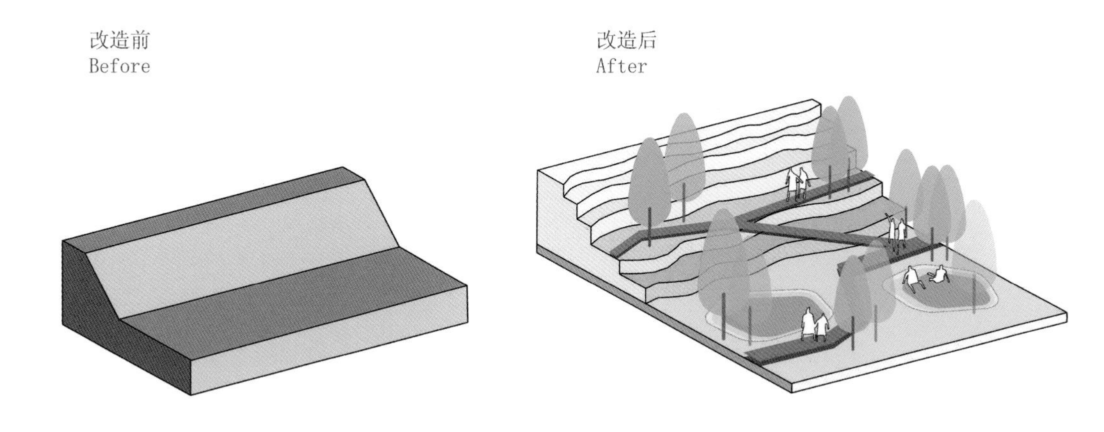

图 2-010 陡坡填土、营造台田策略模式图
Figure 2-010 Steep slope embankment and terraced farmland model diagram

3 规划设计及建成效果
PLANNING & DESIGN AND BUILT EFFECT

3.1 规划方案

PLANNING CONCEPT AND PLAN

3.1.1 设计概念及方案

DESIGN CONCEPT AND PLAN

邯郸园博园的核心设计理念是“上山入水，用地因势；博古通今，溯源启新”。设计通过上述四大生态修复策略给工业废弃地恢复生机，注入活力，使场地实现自然景观的再现与再生，场地在经历了农业文明、工业文明的洗礼后，最后回归完整的生态系统，利用自然的自我调节和净化能力来治愈工业时代留下的城市斑锈。

The core design concept of Handan Garden Expo Park is “Make use of the existing terrain of the mountains and waters, integrate the past and present, trace the origin to inspire the new”. Using the above four ecological restoration strategies, the design restores the industrial wasteland to life and empowers it with vitality, so that the site can reproduce and regenerate natural landscape. After experiencing the baptism of agricultural and industrial civilizations, the site finally returns to a complete ecosystem, and uses nature’s self-regulating and purifying ability to cure the urban rust spot left by the industrial age.

邯郸园博园经济技术指标　　表 3-01

Economic index of Handan Garden Expo Park　　Table 3-01

序号 No.	名称 name	面积 area/m^2	百分比 percentage/%
1	绿地 green space	1428722	50.5
2	建筑 building	70119	2.5
3	铺装 pavement	207000	7.3
4	道路 road	234244	8.3
5	停车场 parking	37140	1.3
6	湿地 wetland	510245	18.1
7	村落 village	340790	12
总计 total		2828230	100

图 3-001 总平面图
Figure 3-001 Master Plan

3.1.2 功能分区规划

FUNCTIONAL ZONING

全园规划了核心文化游览区、生态修复实践区、自然湿地保护区和地域特色展示区四大特色分区，以及齐村民俗体验区、涧沟村农旅文化区和特色工厂改造区三大配套分区。

园博园核心文化游览区位于齐村大坝东侧，包括主入口、山水邯郸主场馆、赵都新韵、地市展园、工业遗风、浮光揽月和青山画卷等核心景观节点，集中展现园林艺术内涵；北侧为地域特色展示区，以展示邯郸文化的民俗印象、体现新型农业的果香农趣园以及母子乐园组成；西南侧为生态修复实践区，包括具有中水净化和科普展示功能的清渠如许、棕地修复的矿坑花园、地市展园以及涧沟陈展馆；西侧的自然湿地保护区保留了西湖水域一片环境良好的自然湿地，充分利用最小干预设计理念，打造观光游览空间。

Four characteristic areas and three supporting areas were planned in the park, with the former including a core cultural tourism area, an ecological restoration practice area, a natural wetland protection area and a regional characteristic exhibition area, and the latter including a folk culture area, Jiangou agricultural culture area and factory rehabilitation area.

The core cultural tourism area of the park is located on the east side of the Qicun Dam. Key landscape nodes within this area include Main Entrance, Main Hall of Handan Pavilion, New Handan Pavilion, City Exhibition Garden, Industrial Heritage Pavilion, Floating Moon Lake, Flowering Terraces etc., showcasing the connotation of garden art. On the north side locates the regional characteristic exhibition area, which is composed of the Folk Culture Pavilion, the Orchard Park, and the Family Park. In the southwest side is the ecological restoration practice area, including Wetland Terraces with functions of reclaimed water purification and science popularization, The Reclaimed Quarry Park for brownfield restoration, the City Exhibition Garden and the Jiangou Exhibition Pavilion. The natural wetland protection area on the west side retains a natural wetland of the Xihu Lake, making full use of the minimal intervention design concept.

除此之外，东侧为齐村民俗体验区，展示新农村建设风貌及部分配套功能；东南侧的涧沟村农旅文化区依托涧沟村悠久的历史文化和涧沟遗址，通过原生村落形态和周边农田风光，形成特色农旅村庄的游园产品；南侧钢城农趣服务区为特色工厂改造区，展示及售卖当地特色农业产品，并为园区提供配套服务。

七大分区特色鲜明，互相配合，为游客提供多元化、多角度、多空间的游览体验。

In addition, on the east side lies the folk custom experience area of Qi Village, which showcases the new countryside with some of its supporting functions. The Jiangou Village Agricultural Tourism Cultural Area on the southeast side relies on the long history and culture of Jiangou Village and the Jiangou Ruins, which form the touring products of characteristic agricultural and tourism villages based on the original village and farmland scenery. The Steel City Farm–Fun Service Area on the south side is a characteristic factory transformation area, which exhibits and sells local agricultural products and provides supporting services.

These seven areas with distinctive features cooperate with each other to provide tourists with a diversified, multi–angle and multi–space experience.

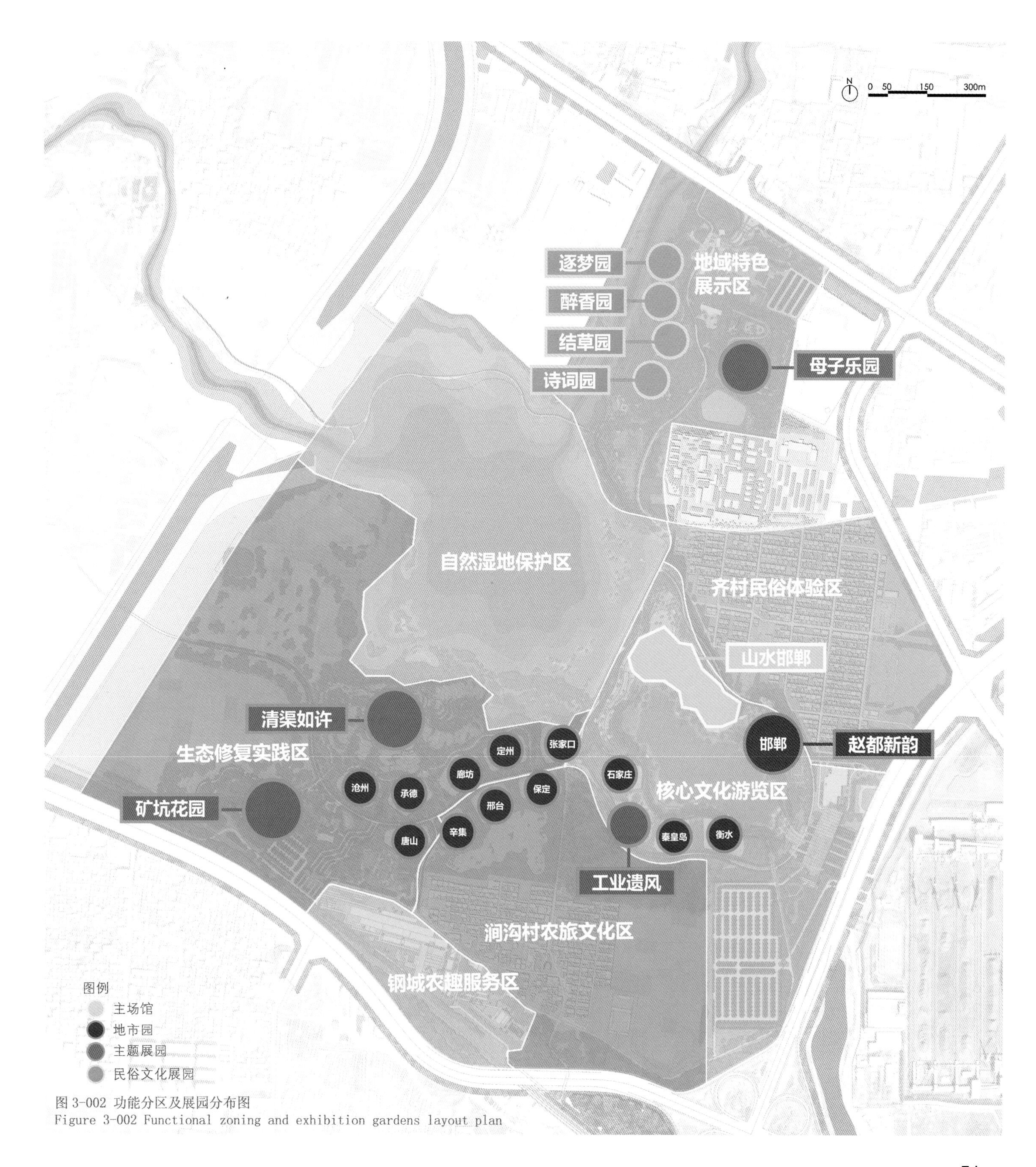

图 3-002 功能分区及展园分布图

Figure 3-002 Functional zoning and exhibition gardens layout plan

3.1.3 交通系统规划

TRANSPORT PLANNING

邯郸园博园的道路设计遵循慢行为主、车行为辅的原则，依据安全、网络化和多维化的原则，合理布局交通体系， 包括车行与慢行系统的设计、出入口及停车场的设计、游览线路与后勤保障线路的设计等，同时满足会展期间和会展后不同的游线需求。

The road design of Handan Garden Expo Park takes slow traffic first, and vehicle traffic as auxiliary. The traffic system is arranged in line with safety, networking and multi-dimensionality requirements, including the design of vehicle traffic and slow traffic system, entrances, exits and parking space, tour routes and logistical support routes, etc., while meeting needs during and after the expo.

全园共设置 3 个主要人行出入口，分别为东入口、南入口和北入口，供开园时游客通行。其中，东入口为园博园主入口，紧邻西环路；北入口紧邻新邯武公路；南入口临近邯武快速路。全园共设 7 个车行出入口，供管理车辆和后勤车辆通行使用，其中主入口 2 处、南入口 1 处、北入口 1 处、西入口 1 处，齐村和涧沟村入口各 1 处。对外大型停车场共 3 处，分别位于东、南、北 3 个主要出入口，酒店、涧沟陈展馆等场馆各自布置内部停车场。

There are three main pedestrian entrances in the park, respectively located in the east, south and north. Among them, the main entrance in the east adjoins the West Ring Road. The north entrance is adjacent to the New Handan–Wu'an Highway. The south entrance is adjacent to the Handan–Wu'an Expressway. There are a total of seven vehicle entrance and exit gates for management and logistics vehicles, including two main entrances, and one each respectively in south, north, west, and in Qi and Jiangou villages. There are three large–scale external parking lots, located at the three main entrances in the east, south, and north, respectively. The hotel, the Jiangou Museum and other venues are equipped with internal parking.

邯郸园博园停车场统计表
Statistical table of parking space

表 3-02
Table 3-02

	红线内 within redline		红线外（临时）out of redline (temp.)		总计 total
	小型客车（辆）light–duty vehicle	大型客车（辆）bus / coach	小型客车（辆）light–duty vehicle	大型客车（辆）bus / coach	
主入口 main entrance	1130	—	—	—	1130
北入口 north entrance	216	—	1033	56	1308
南入口 south entrance	172	—	—	—	172
涧沟博物馆 Jiangou Museum	36	—	—	—	36
总计 total	1554	—	1033	59	2646

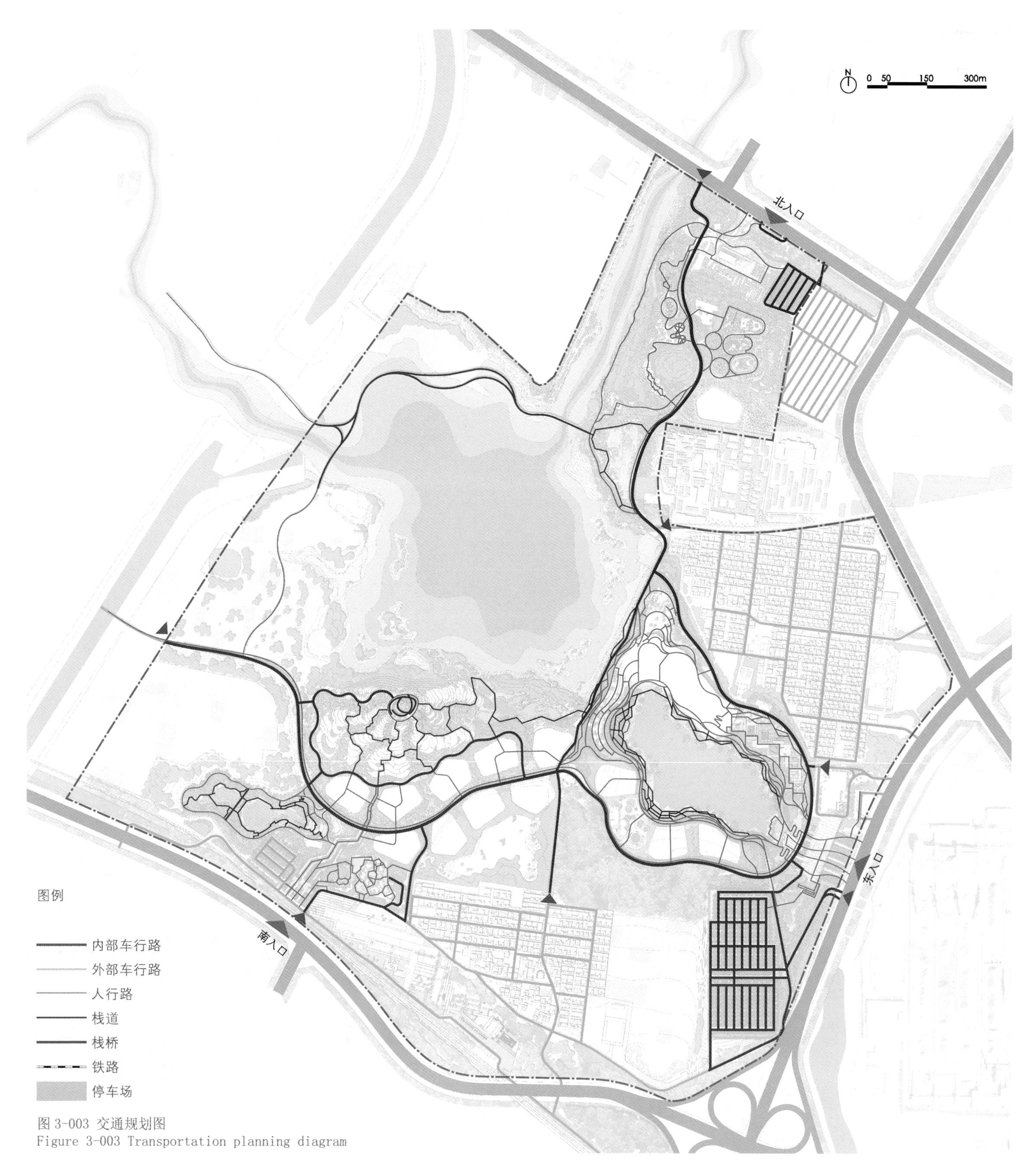

图 3-003 交通规划图
Figure 3-003 Transportation planning diagram

（1）车行交通

VEHICULAR TRAFFIC

设计依托现状道路，并根据场地原有地形，自北入口蜿蜒延伸，经齐村大坝堤顶路与主入口连接。核心文化游览区外围的车行主路与齐村大坝南北两端相接，形成园区内部环路，整体构成“一带一环”的车行系统，并与各出入口相接。园区内部的人行道保证了所有场地的通行可达性，并形成便捷的慢行网络，为游客提供便利的游览系统，丰富其游园感受。

The design of vehicle roads relies on the existing road and the original topography of the site. Winding and extending from the north entrance, it is connected to the main entrance via the top road of the Qicun Dam. The main road on the periphery of the core tourism area connects with the north and south ends of the Qicun Dam, forming an inner loop of the park. The vehicular road system constitutes a belt and a ring, and is connected with each entrance. The walkways inside the park ensure the accessibility of all venues and form a convenient slow-traffic network to provide visitors with a convenient tour system.

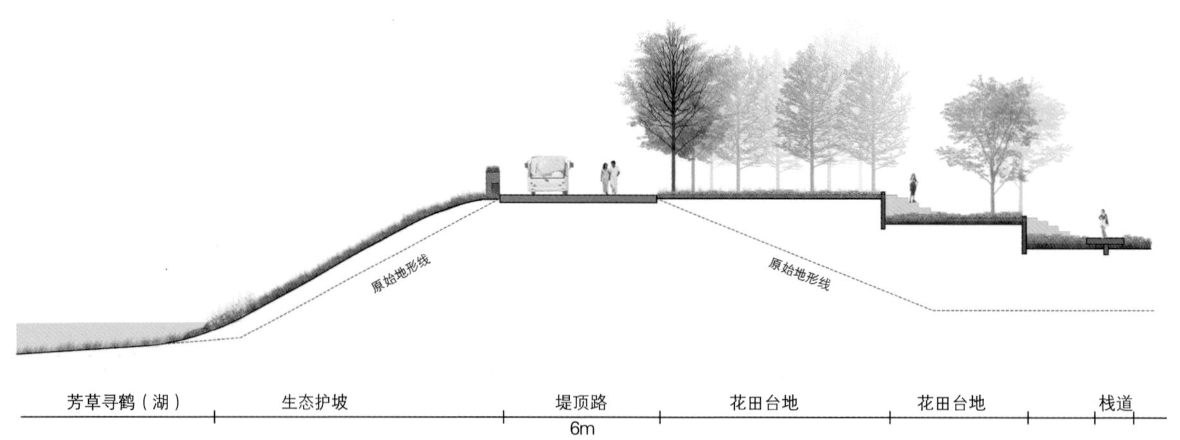

图 3-004 车行道剖面图（齐村大坝）
Figure 3-004 Driveway section (Qicun Dam)

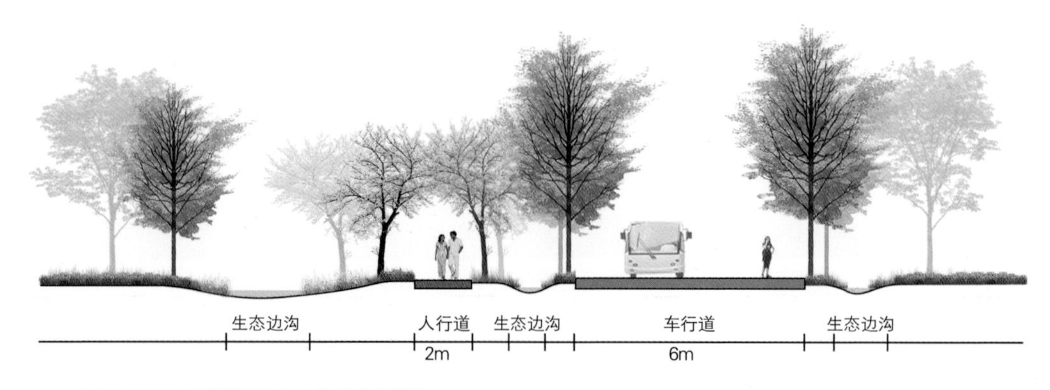

图 3-005 车行道剖面图（单侧步行道）
Figure 3-005 Driveway section (sidewalk on one side)

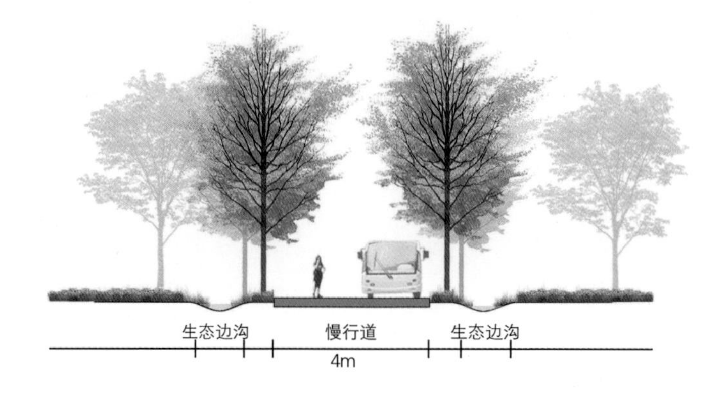

图 3-006 车行道剖面图（次园路）
Figure 3-006 Driveway section (secondary park road)

图3-007 齐村大坝建成实景
Figure 3-007 Built photo of the Qicun Dam

（2）人行交通
PEDESTRAIN CIRCULATION

人行道包括1.5m、2m、3m和4m四个级别，充分考虑无障碍设计，形成完善的慢行网络。

The walkways include four levels of 1.5m, 2m, 3m and 4m, with full consideration of barrier-free design, forming a thorough slow-traffic network.

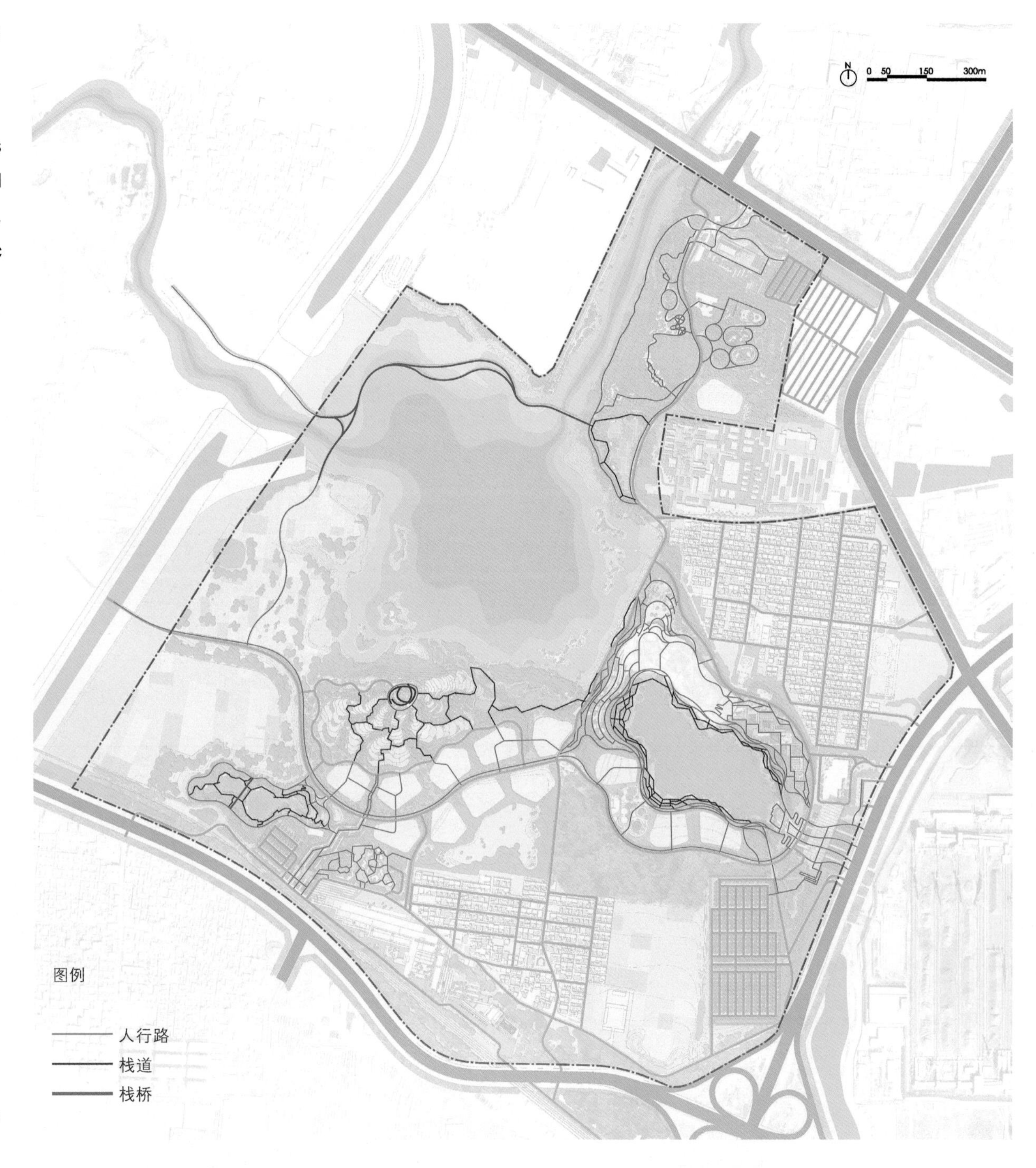

图3-008 人行交通分析图
Figure 3-008 Pedestrian circulation analysis

图 3-009 人行道建成实景
Figure 3-009 Photo showing sidewalk upon construction

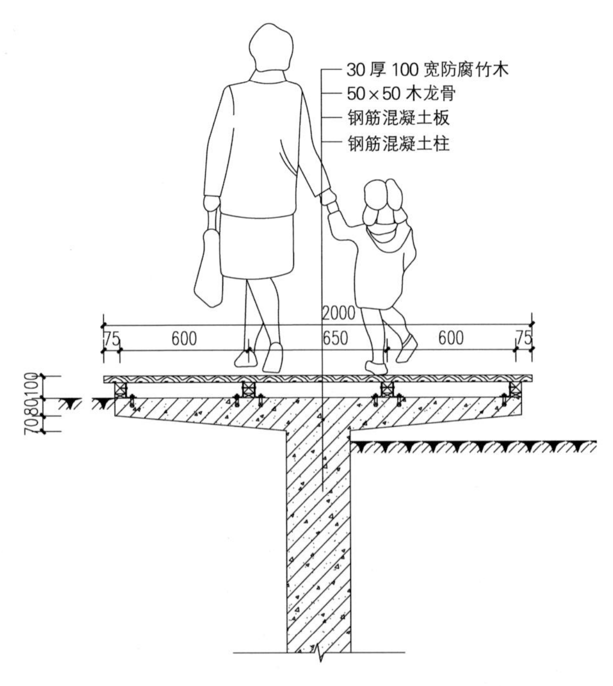

图3-010 2m木栈道剖面图
Figure 3-010 2m Wooden boardwalk section

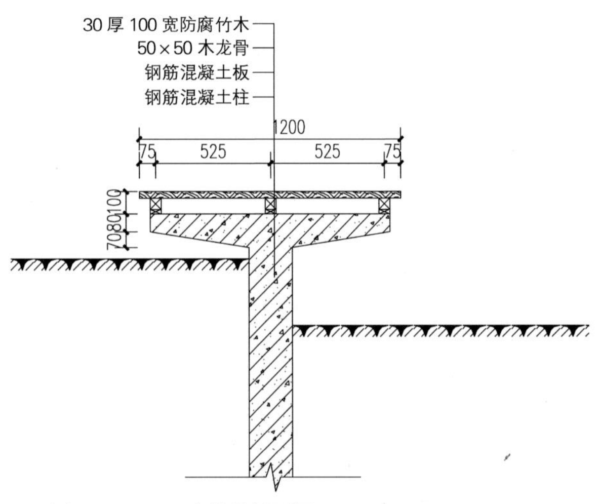

图3-011 1.2m木栈道剖面图
Figure 3-011 1.2m Wooden boardwalk section

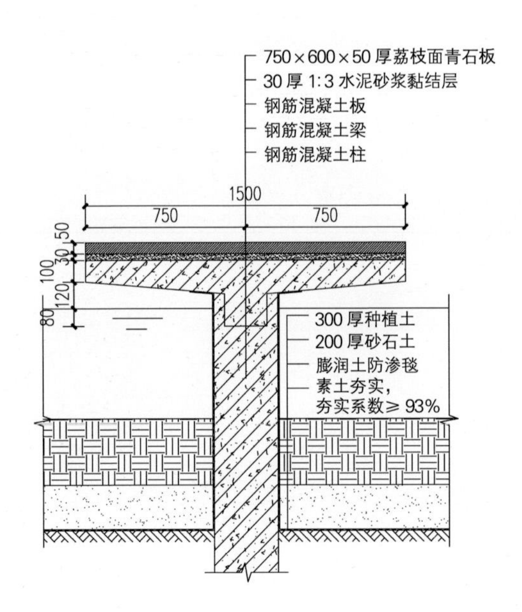

图3-012 1.5m石材栈道剖面图
Figure 3-012 1.5m Wooden boardwalk section

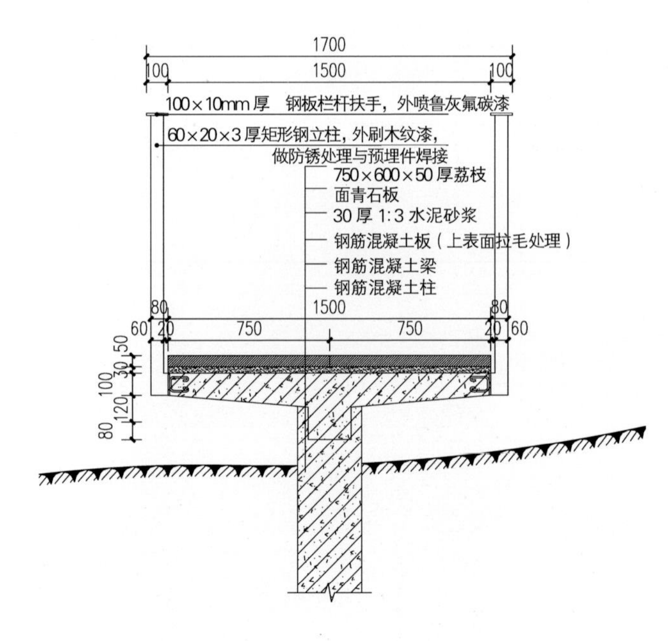

图3-013 1.7m石材栈道剖面图
Figure 3-013 1.7m Wooden boardwalk section

图 3-014 木栈道施工过程实景（浮光揽月）
Figure 3-014 Photo showing wooden boardwalk under construction (at Floating Moon Lake)

图 3-015 石材栈道施工过程实景（矿坑花园）
Figure 3-015 Photo showing wooden boardwalk dock under construction (at Reclaimed Quarry Park)

图 3-016 人行栈道
Figure 3-016 Pedestrian boardwalk

3.1.4 水系规划

WATER SYSTEM PLANNING

园区内现状水体为西湖水库，设计的滨水景观节点包括浮光揽月、清渠如许以及矿坑花园等。

西湖水库位于沁河末端，被齐村大坝阻隔，设计利用涵管将浮光揽月与西湖水库连通，打造园区内核心景观水体，雨季时溢出，流入外部水体。

园区内还设计了中水净化展示节点——清渠如许。将中水通过提水泵引水至清渠如许山顶，经过人工湿地净化后的中水用作园区内景观用水，并部分回用于植物浇灌、冲厕、市政浇洒等。

矿坑花园水体与清渠如许连通，经由清渠如许人工湿地净化后的中水汇入矿坑花园水系景观，最终由矿坑花园溢流汇入西湖水库。

The current water body in the park is the Xihu Dam. Designed waterfront landscape nodes include Floating Moon Lake, Wetland Terraces, and The Reclaimed Quarry Park.

The Xihu Dam is located at the end of the Qin River and is blocked by the Qicun Dam. The design uses culverts to connect the Floating Moon Lake to the Xihu Dam to create the core landscape water body in the park, which overflows into the external water body during rainy seasons.

A reclaimed water purification exhibition node was designed in the park—The Wetland Terraces. The reclaimed water is diverted to the top of the Wetland Terraces mount through a water lift pump. After purification by the wetland, the reclaimed water is used for landscape water body, plant irrigation, toilet flushing, and municipal watering.

The water body of The Reclaimed Quarry Park is connected with the Wetland Terraces. The purified reclaimed water flows into the water system of The Reclaimed Quarry Park, and finally overflows into the Xihu Dam.

全园四大景观水体相互连通，形成了两条净化路线。

①净水路线 1：中水补水点—清渠如许西侧 2 条潜流—表流复合湿地—内河湿地—矿坑花园水体—西湖水库。

②净水路线 2：中水补水点—清渠如许东侧 3 条潜流—表流复合湿地—内河湿地—西湖水库。

The four major landscape water bodies within the park are connected to each other, forming two purification routes.

① Water purification route 1: Reclaimed water replenishment point—2 subsurface flow compound wetland west of Wetland Terraces—surface flow wetland—inland river wetland—The Reclaimed Quarry Park water body—Xihu Dam.

② Water purification route 2: Reclaimed water replenishment point—3 subsurface flow compound wetland—surface flow wetland—inland river wetland—Xihu Dam.

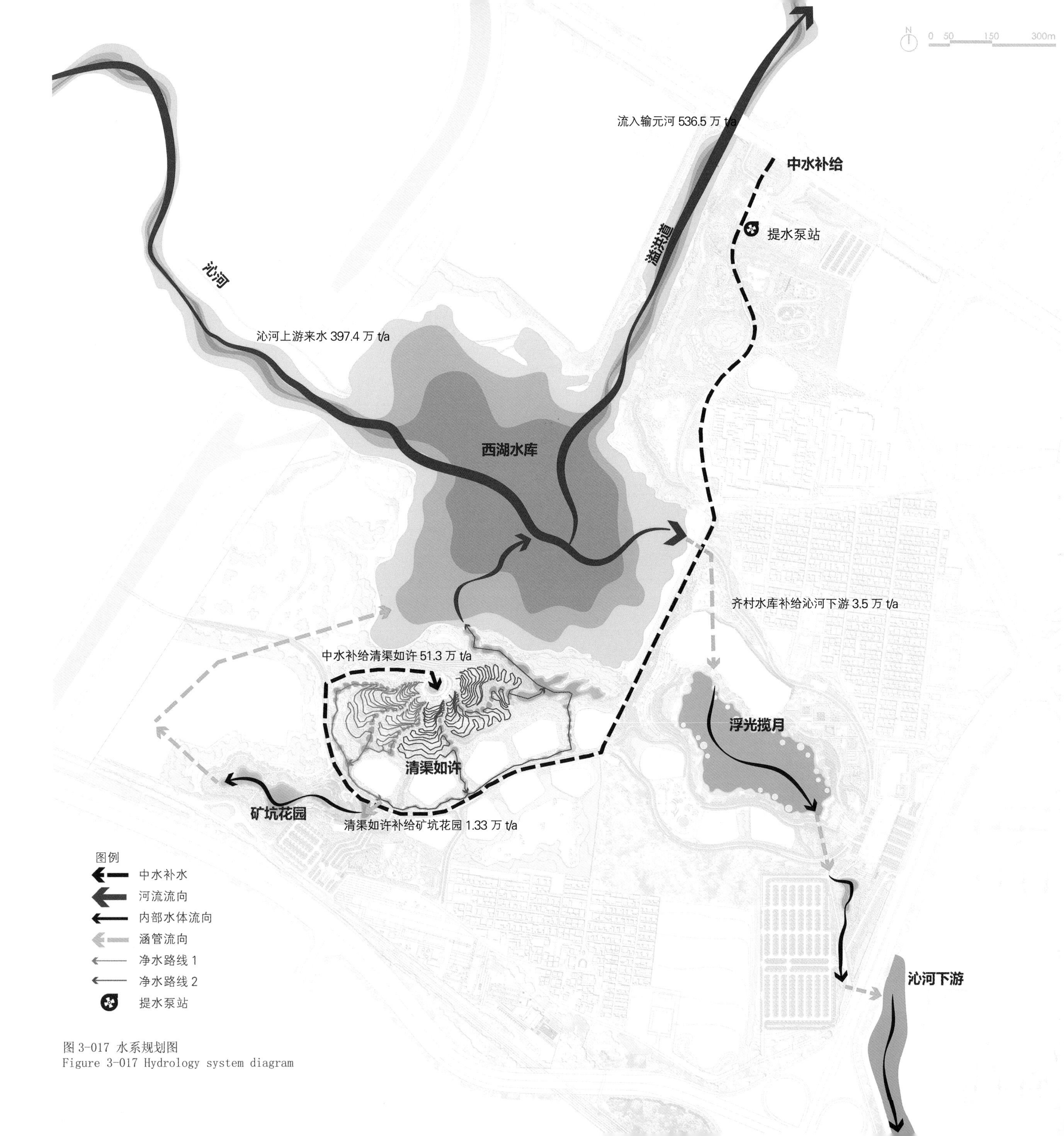

图 3-017 水系规划图
Figure 3-017 Hydrology system diagram

3.2 节点设计

Nodes Design

以“一条花路串园博十景”形成整个园区的景观结构，从核心文化游览区的浮光揽月、赵都新韵、山水邯郸、青山画卷、工业遗风到生态修复实践区的清渠如许和矿坑花园，以及自然湿地保护区的梦泽飞虹、邯郸印塔，地域特色展示区的民俗印象等，共同形成完整的景观意象。

本节基于园博园的游览顺序，从3个出入口以及园博十景来分别诠释邯郸园博园的重要景观设计理念。

The landscape structure of the entire park is formed under the concept of “Ten scenic nodes connected by one flower road”. Floating Moon Lake, New Handan Pavilion, Handan Pavilion, Flowering Terraces, Industrial Heritage Pavilion in the core cultural tourism area, Wetland Terraces and Reclaimed Quarry Park in the ecological restoration practice area, and the Flying Bridge, Handan Tower in the natural wetland reserve, and Folk Culture Pavilion in the regional characteristic exhibition area, all together form a complete landscape image.

Based on the tour sequence, this chapter interprets the important landscape design concepts of the park from the three entrances and the ten scenic spots.

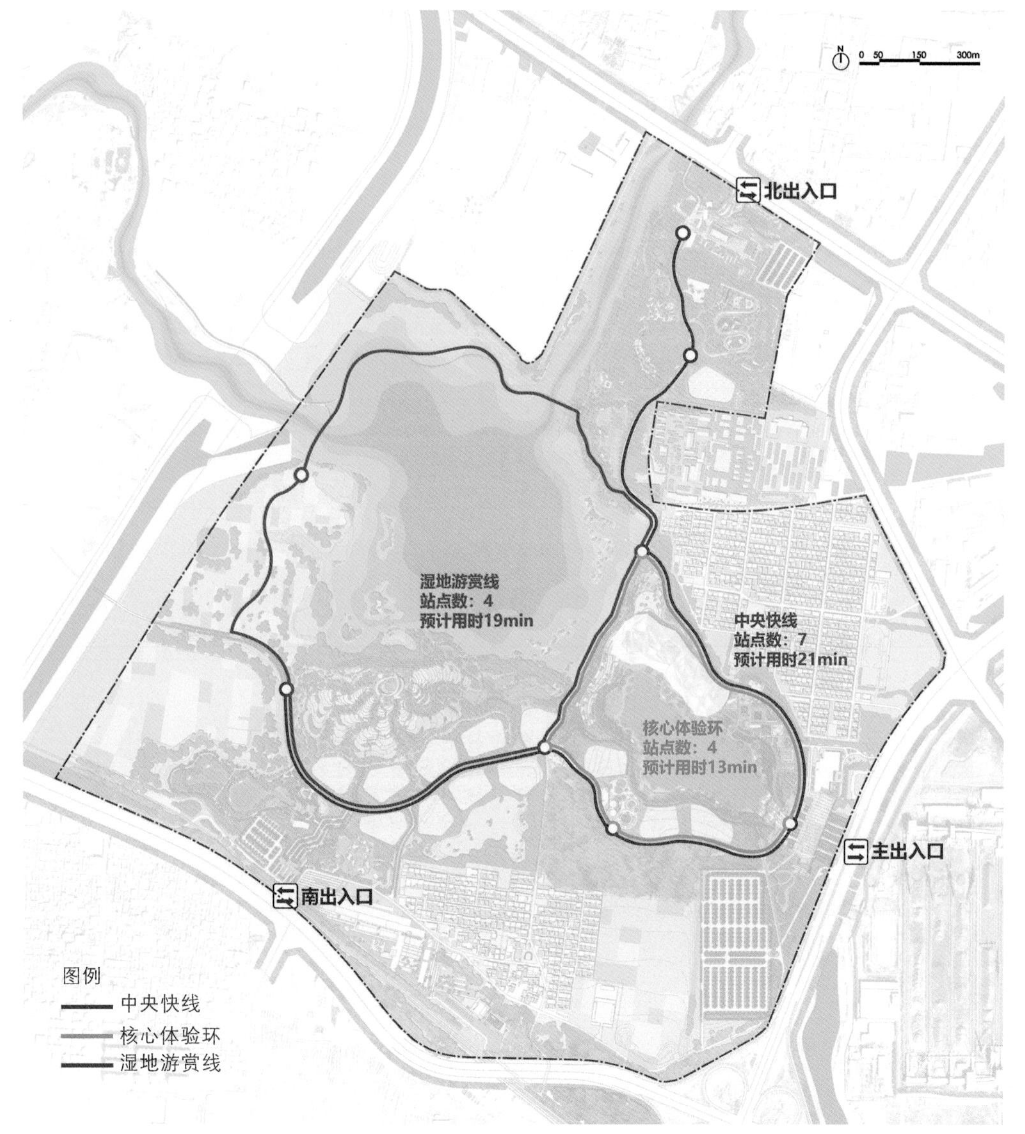

图 3-018 电瓶车游线规图
Figure 3-018 Electric vehicle circulation plan

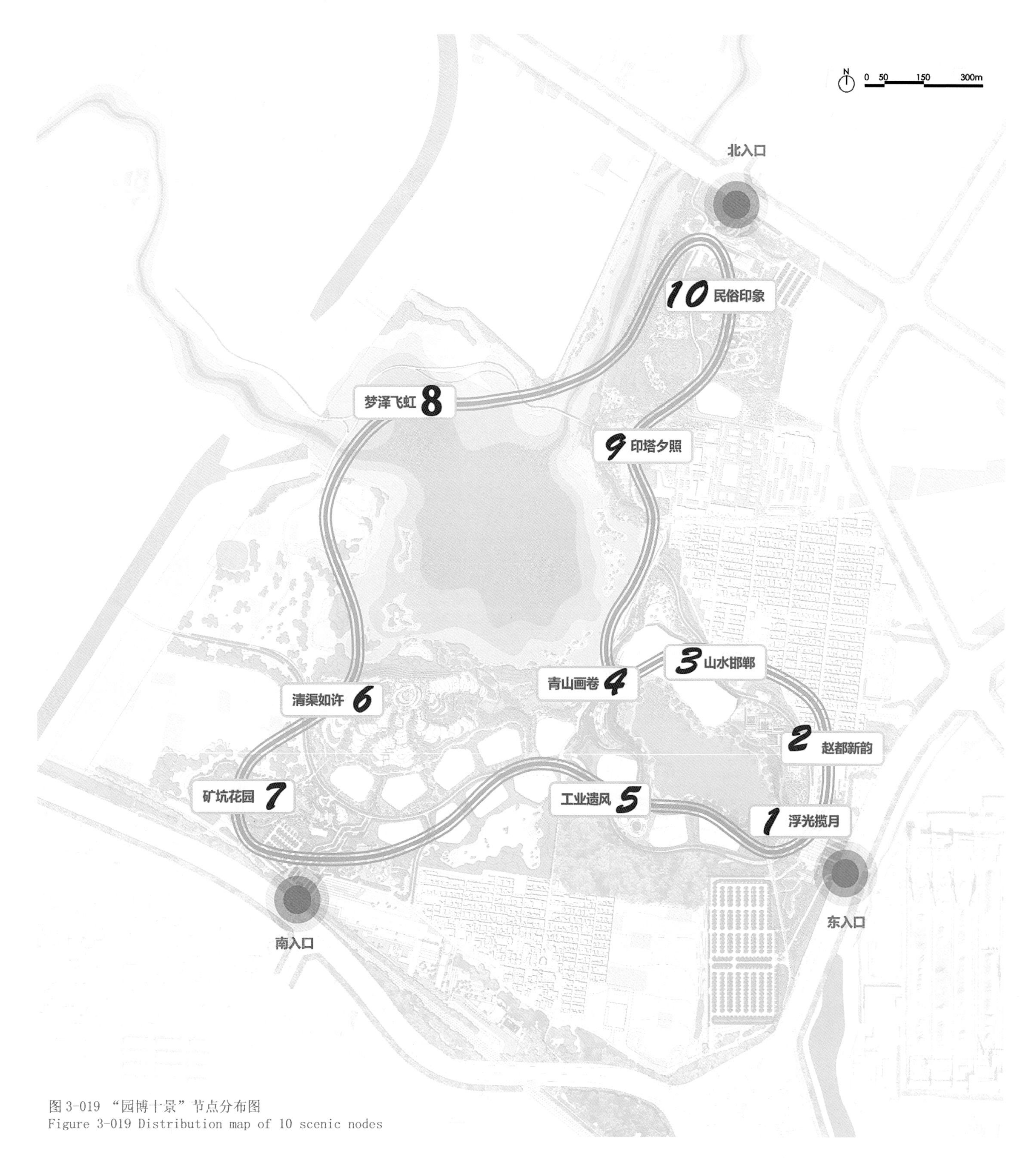

图 3-019 “园博十景”节点分布图
Figure 3-019 Distribution map of 10 scenic nodes

3.2.1 东入口
EAST ENTRANCE

东入口是园博园的主入口，紧邻西环路，是城市与园区的联系界面。作为全园的主要集散场所，场地依托现状平坦的地形地貌，打造开敞的林下广场空间。

整体设计以行云流水为概念，蜿蜒的铺装线条体现沁河复流的设计理念。入口前广场绿树成荫，随着树阵和铺装的引导进入园博园区域后，视线豁然开朗， 12m 高、140m 宽的白色云顶大门屹立眼前，大门舒展开敞，顶部为玻璃钢编织纹样，其丰富的光影变化给人以绚丽夺目的感受，围绕白色钢柱设置异形红色玻璃钢座椅，作为观赏小品的同时为游客提供遮阴功能，整个大门给人留下强烈而深刻的印象。

The east entrance is the main entrance of the park. Adjoining the West Ring Road, it is the interface between the park and the city. As the main gathering and distributing place of the whole park, it relies on the current flat terrain to create an open plaza space under the canopy.

The overall design tends to follow a smooth and natural concept. The meandering paving lines reflect the concept of the reflow of Qin River. The square in front of the entrance is lined with trees. After entering the park guided by the trees and the paving, the line of sight suddenly opens up. The 12-meter-high and 140-meter-wide white cloud-top gate stands in front, the top of the gate is made into the glass fabric reinforced plastics texture, with its rich light and shadow changes giving visitors a gorgeous and eye-catching feeling. Irregular-shaped red glass fiber seats are set around the white steel pillar, which are used as ornamental sketches and provide shade for tourists. The entire gate makes a strong and deep impression.

图 3-020 东入口
Figure 3-020 East entrance

图3-021 东入口大门
Figure 3-021 East entrance gate

图 3-022 红色玻璃钢坐凳
Figure 3-022 Red FRP benches

3.2.2 北入口

NORTH ENTRANCE

北入口广场是对原水泥厂厂房的改造，设计以“绿色脉动”为主题，同样采用流线的设计语言，消弭场地的高差变化，并形成了强烈的引导性，便于人流快速疏散。

由新邯武公路到达北入口广场，首先映入眼帘的是顺应场地高差变化设计的18级台阶及其后的标志性大门构筑物。大门由两片错位摆放的锈蚀钢板墙体组成，分别列于视线的正前方与左侧，墙体运用镂空的设计手法，呈现邯郸城市剪影的图案，以增加层次感和空间感，背面的图案是云影，与正面图案互相呼应。夜晚景墙内的灯光也会透过镂空图案倾洒在地面上，形成别具一格的设计效果。

The north entrance plaza is a renovation of the former cement factory. The design is based on the theme of “green pulse”. It adopts a streamlined design language to eliminate the height difference of the site and form a strong guide to facilitate the flow of people during the expo.

Arriving from the New Handan–Wu’an Highway to the north entrance plaza, the first thing that catches one’s eyes is the 18 steps and the iconic gate structure followed. The gate is composed of two staggered rusty steel plate walls, one of which directly in front and the other on the left side of the line of sight. Hollow design techniques are used in the walls to present the pattern of the silhouette of Handan City, as well as increase the sense of spatial dimensions. The pattern of cloud shadow on the backside echoes with the front side. At night, light will be poured on the ground through the hollow pattern, forming a unique design effect.

图 3-023 北入口大门构筑物
Figure 3-023 North entrance gate

图 3-024 北入口全景
Figure 3-024 Panaroma of north entrance

图 3-025 北入口广场
Figure 3-025 North entrance plaza

图 3-026 北入口餐厅
Figure 3-026 Canteen near north entrance

正前方为原有厂房建筑改造的餐厅。新建建筑以极为纯净的玻璃体嵌入原有厂房，新旧建筑体之间形成强烈的材料对比，又达到和谐统一。新植入的建筑体量悬浮在工业厂房结构骨架的中间，首层四周采用可开启的连续玻璃门扇与周边景观无缝衔接，打造高品质的观赏空间。建筑下方设置倒影水池，将建筑及周边景观映像其中，美轮美奂，丰富视觉感受，体现了场地新旧记忆的完美融合，成为北入口的标志物。

Directly in the front is the restaurant reconstructed from a factory building. A newly constructed building, with extremely pure glass, is embedded in the existing factory building, forming a strong material contrast between the old and the new, while achieving harmony. The newly implanted building is suspended in the middle of the structure of the industrial plant. A continuous series of openable glass doors are used around the first floor to seamlessly connect with the surroundings, creating a high-quality viewing space. A reflection pool is set under the building to it and the surrounding landscape. Magnificently enriching the visual experience, it reflects the perfect fusion of the old and new memories of the site, and becomes the landmark of the north entrance.

图 3-027 北入口原厂房建筑
Figure 3-027 Former factory building near north entrance

图 3-028 夕阳下的北入口餐厅
Figure 3-028 North entrance canteen in the dusk

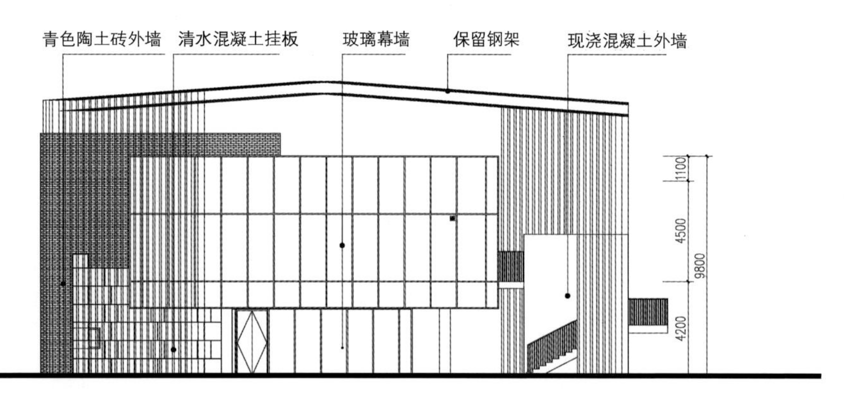

图 3-029 北入口餐厅东立面图
Figure 3-029 East elevation of north entrance canteen

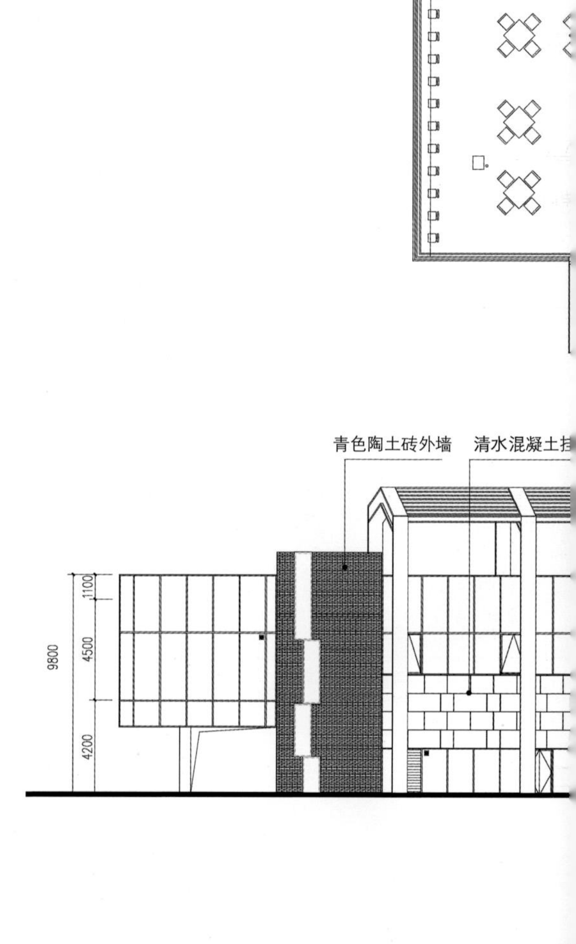

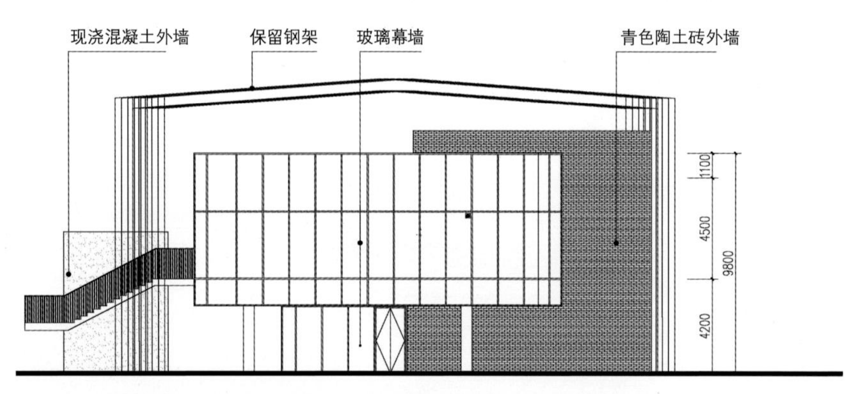

图 3-030 北入口餐厅西立面图
Figure 3-030 West elevation of north entrance canteen

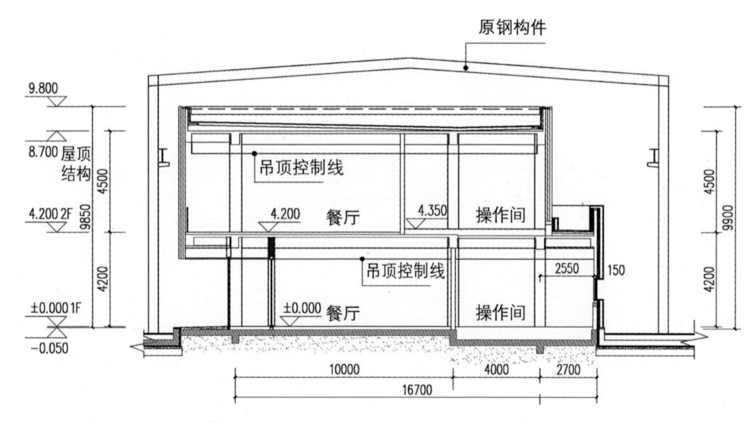

图 3-031 1-1 剖面图
Figure 3-031 Section 1-1

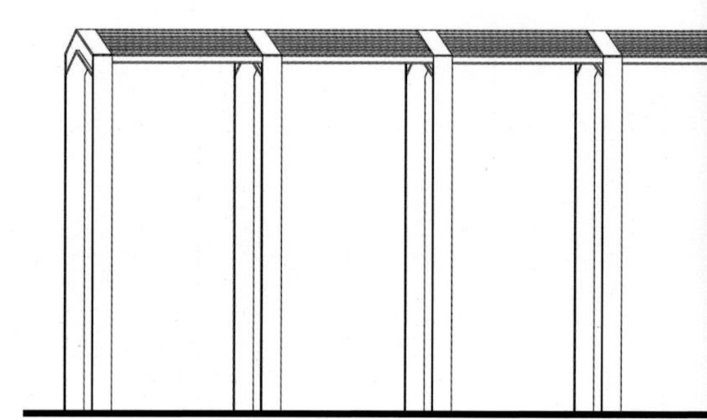

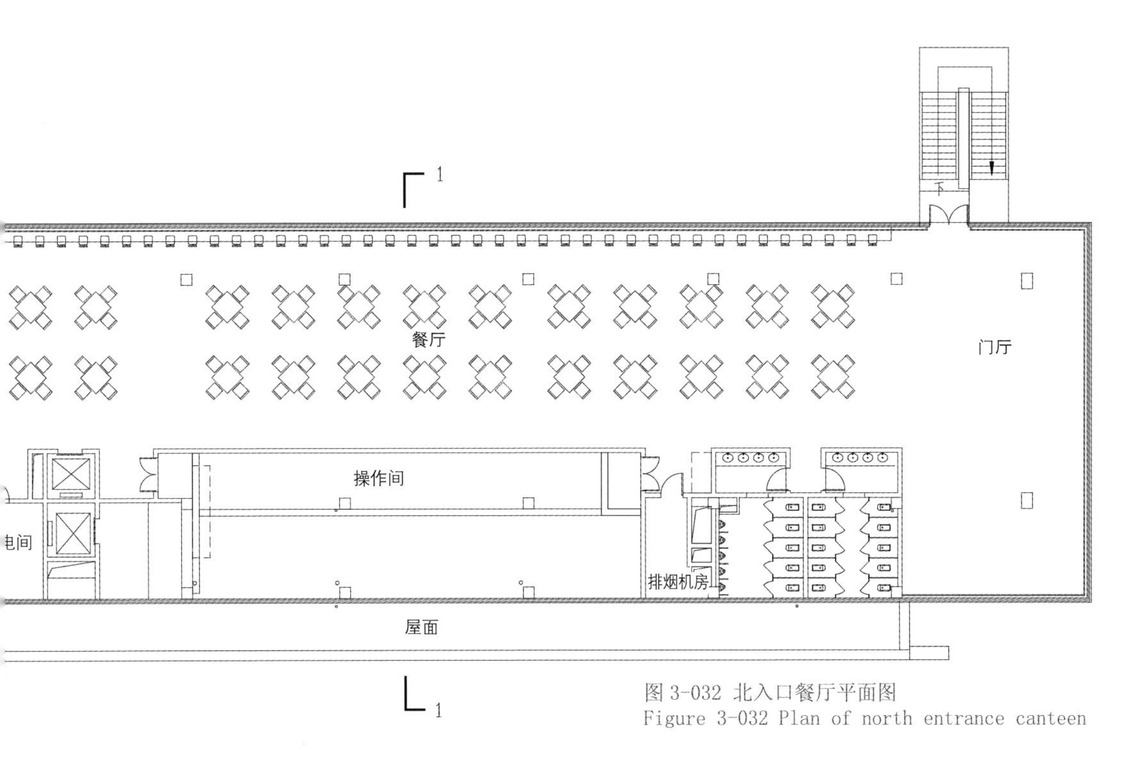

图 3-032 北入口餐厅平面图
Figure 3-032 Plan of north entrance canteen

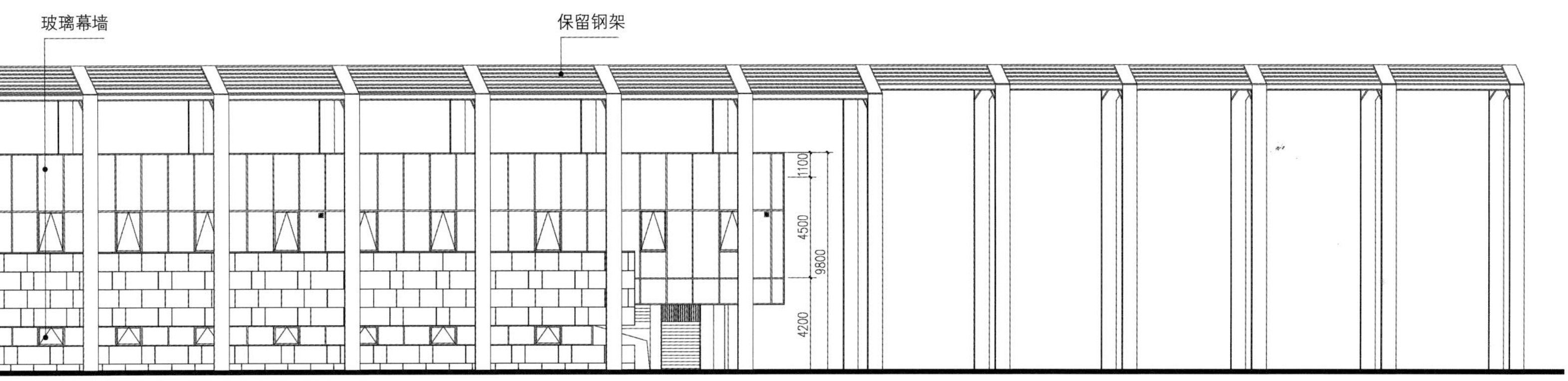

图 3-033 北入口餐厅北立面图
Figure 3-033 North elevation of north entrance canteen

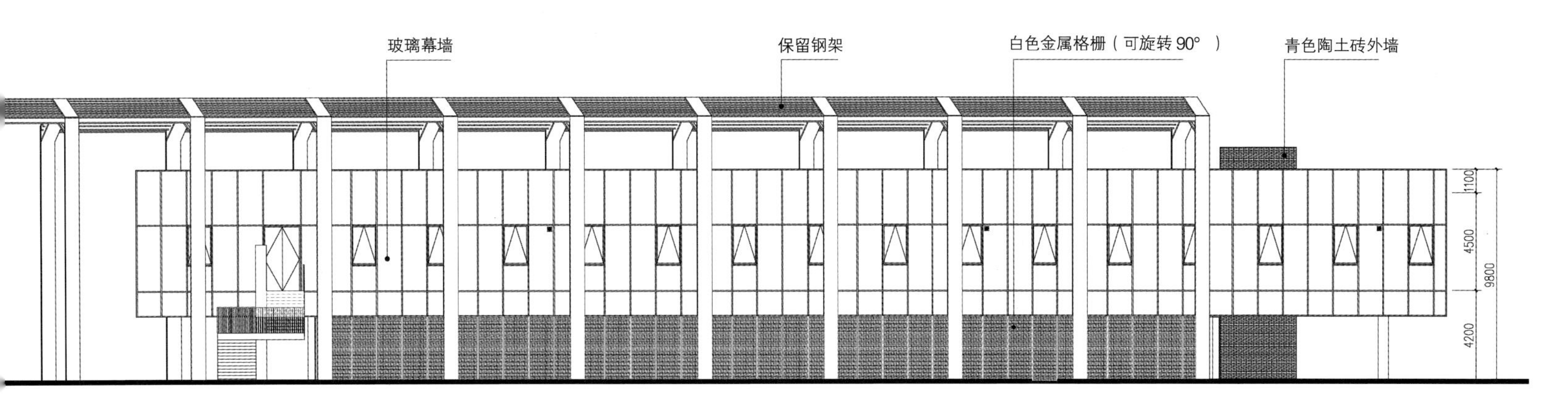

图 3-034 北入口餐厅南立面图
Figure 3-034 South elevation of north entrance canteen

北入口广场的南侧边缘利用台地花田消解高差并保留了水泥厂的3个龙门吊，诉说着场地曾经的工业历史。

The southern edge of the north entrance square uses terraced fields to dissolve the height difference and retains 3 gantry cranes from the cement factory, telling the industrial history of the site.

图 3-035 厂址遗留龙门吊
Figure 3-035 A Gantry-style overhead crane from former factory

图 3-036 龙门吊
Figure 3-036 Gantry-style overhead crane

广场右侧为园区北入口，大门为一组长条形建筑，承担着入口、售票、卫生间、机房、纪念品商店等众多功能，原本零散的功能布局被两道强烈的墙体串联，墙体之间形成的趣味空间，时而为庭院，时而转化为室内空间。

On the right side of the square is the north entrance of the park. The gate is composed of a group of elongated buildings, which are responsible for functions such as entrance, ticket sales, toilet, machine room, souvenir shop, etc. The once dispersed functional layout is now connected by two strong walls. The interesting space formed in between the walls is sometimes a courtyard while other times transformed into an indoor space.

图 3-037 北入口服务中心全景
Figure 3-037 Panaroma of north entrance service centre

图 3-038 北入口服务中心
Figure 3-038 Service centre at north entrance

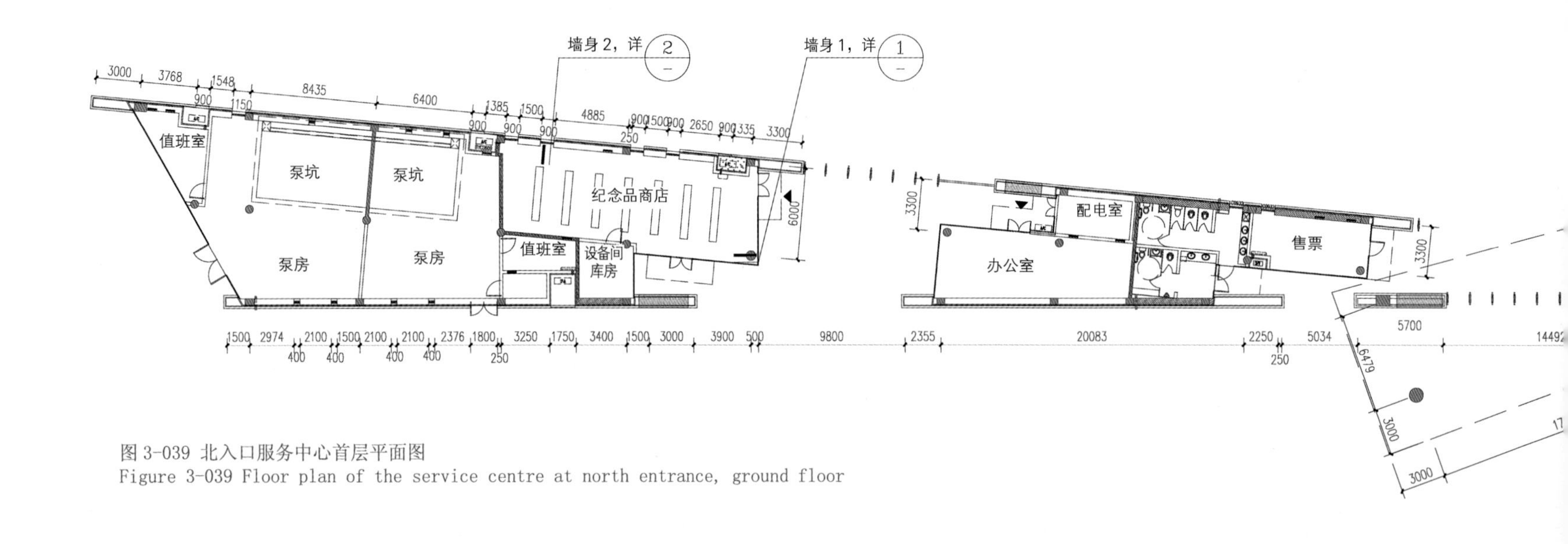

图 3-039 北入口服务中心首层平面图
Figure 3-039 Floor plan of the service centre at north entrance, ground floor

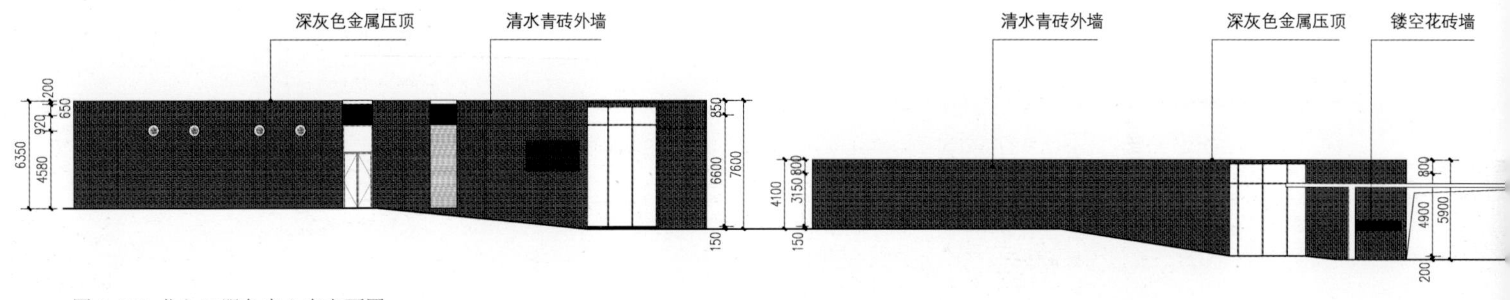

图 3-040 北入口服务中心东立面图
Figure 3-040 East elevation of the service centre at north entrance

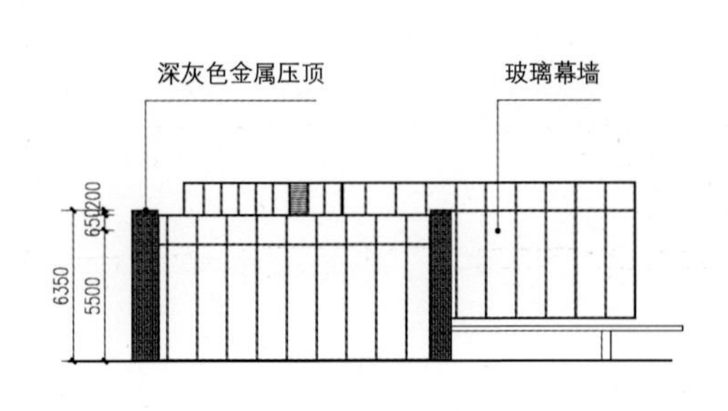

图 3-041 北入口服务中心南立面图
Figure 3-041 South elevation of the service centre at north entrance

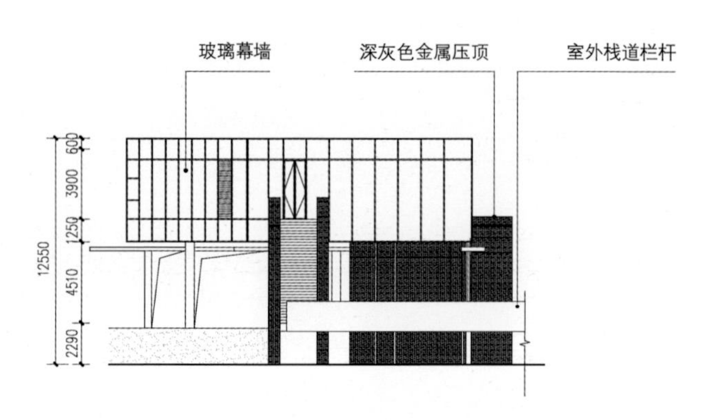

图 3-042 北入口服务中心北立面图
Figure 3-042 North elevation of the service centre at north entrance

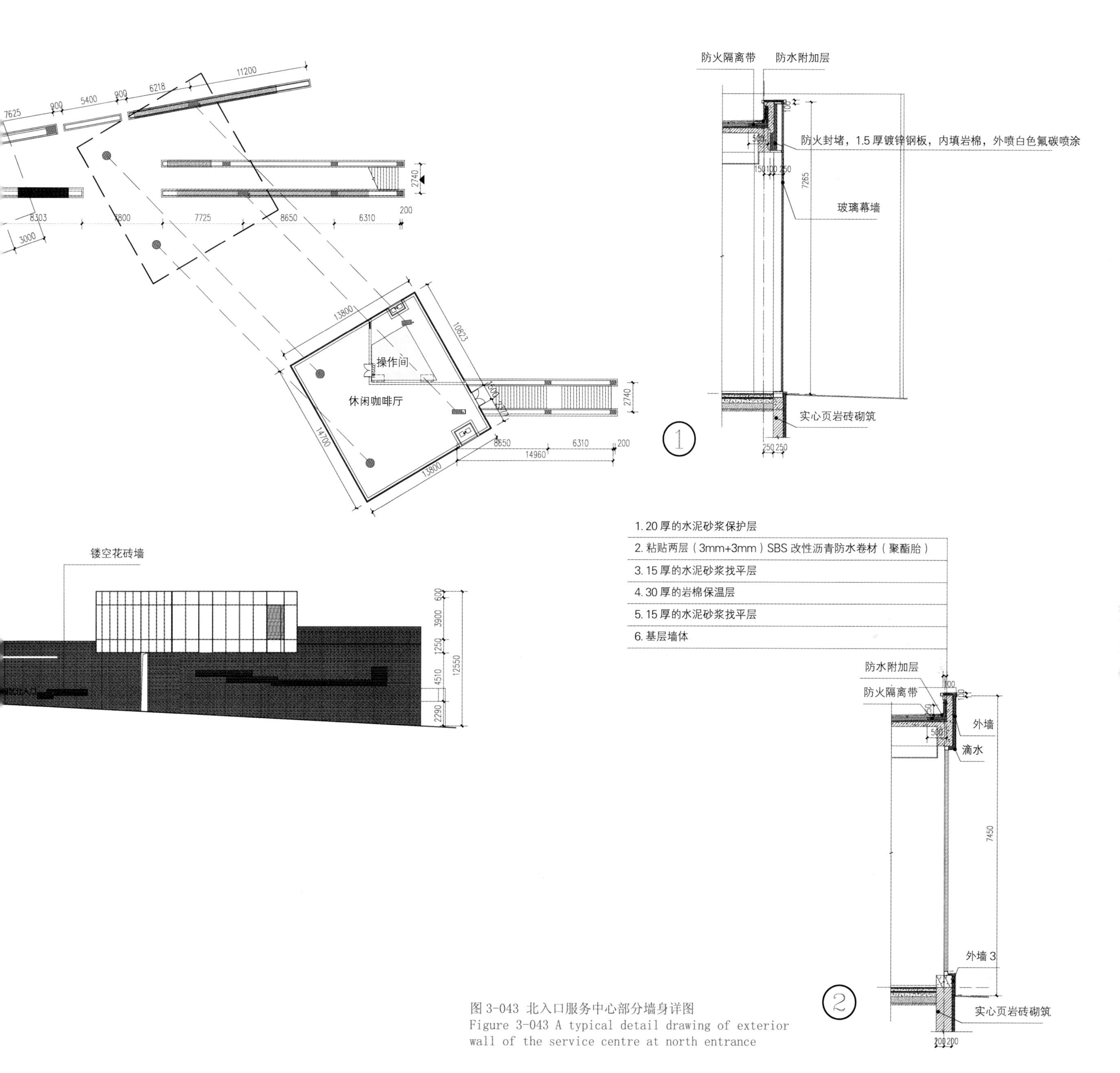

图 3-043 北入口服务中心部分墙身详图

Figure 3-043 A typical detail drawing of exterior wall of the service centre at north entrance

3.2.3 南入口

SOUTH ENTRANCE

南入口临近邯武快速路，入口服务建筑的设计与广场内涧沟陈展馆的体量和材料相呼应，强调区域设计语言的统一，曲折的铺装肌理将游客引导入园区内。

The south entrance is adjacent to the Handan–Wu'an Expressway. The design of the entrance service building echoes the volume and materials of the Jiangou Museum, unifying with the regional design language, which is also used in the winding pavement texture that guides visitors into the park.

图 3-044 南入口全景
Figure 3-044 Panorama of the south entrance

设计采用多边形母题勾勒出连续完整的入口形态，局部开放的庭院空间配合植被生长，空间高低错落，带来丰富的游览体验，形成宜人舒适、特色鲜明的入口，在迎来送往之间尽显诗意情怀。

The design adopts a polygonal motif to outline a continuous and complete entrance form. The partially open courtyard space matches the vegetation growth, making staggered space that forms a pleasant and comfortable entrance with distinctive features, giving poetic feelings to the arrivals and departures.

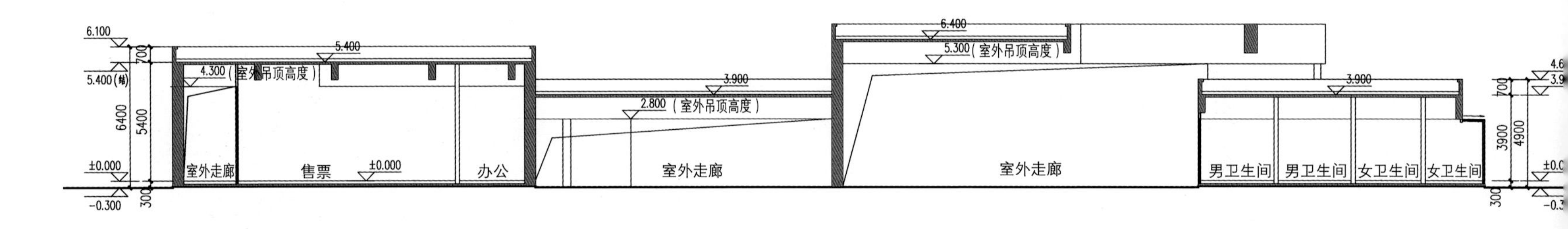

图 3-045 南入口服务建筑剖面图
Figure 3-045 Section of the facility building at south entrance

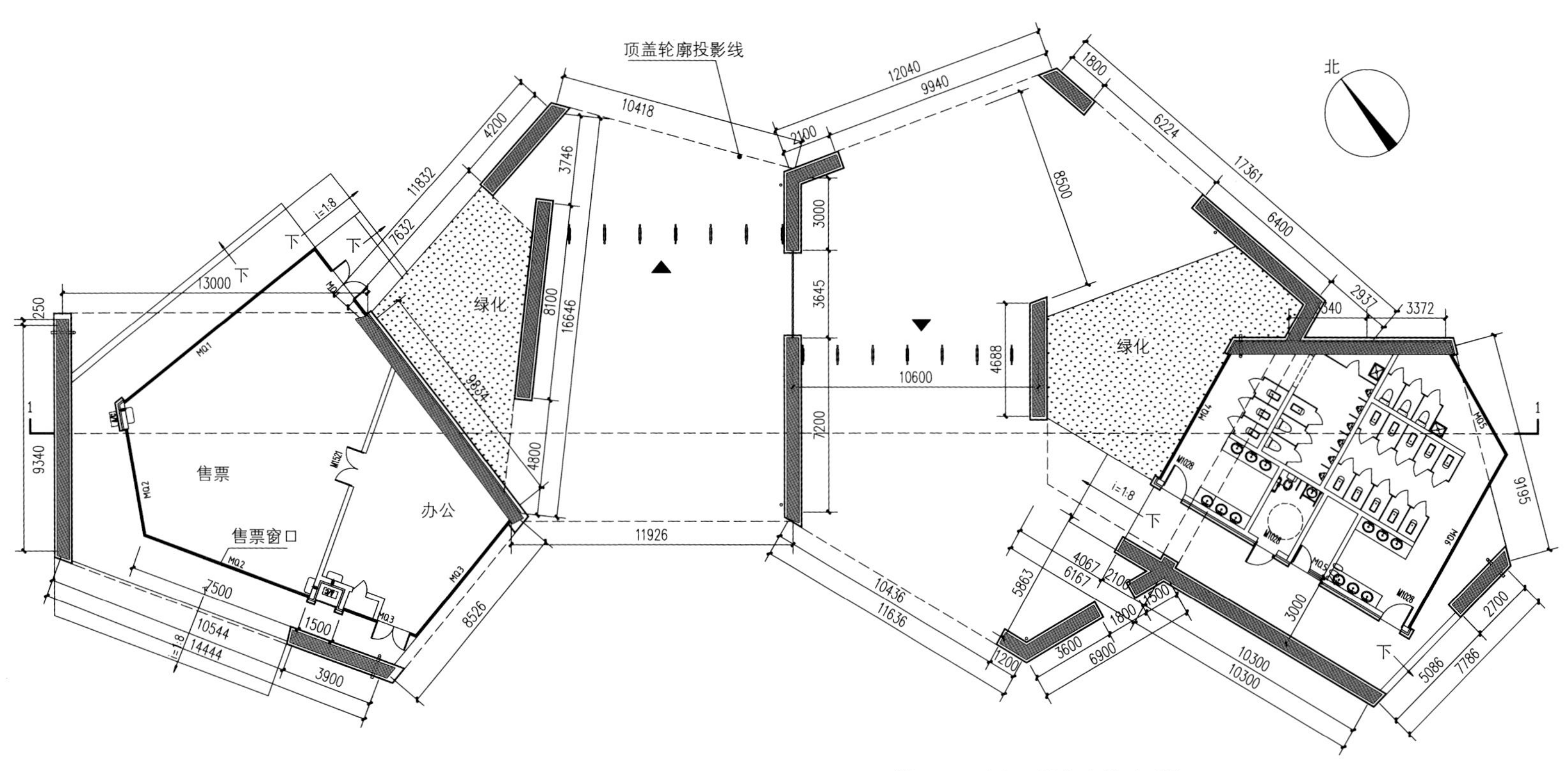

图 3-046 南入口服务建筑平面图
Figure 3-046 Floor plan of the service building at south entrance

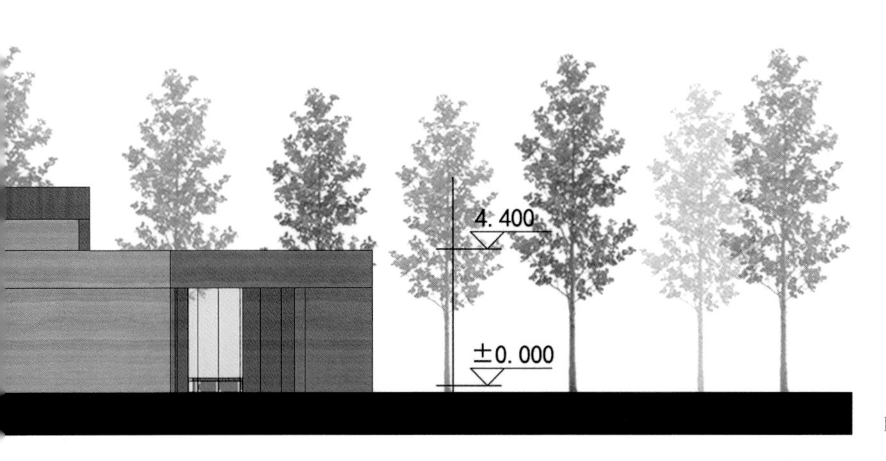

图 3-047 南入口服务建筑正立面图
Figure 3-047 Front elevation of the facility building at south entrance

广场铺装选用不同颜色的长条形花岗石作为主材，在铺装中穿插设置种植池种植乔木，在游人观赏游览的同时起到连续的遮阴作用。

The paving of the square uses long strips of granite of different colors, and planting pools are interspersed in the paving for tree planting, providing continuous shade for visitors.

图 3-048 南入口广场
Figure 3-048 South entrance plaza

图 3-049 南入口广场鸟瞰图
Figure 3-049 Bird eye view of the south entrance plaza

图 3-050 涧沟陈展馆全景
Figure 3-050 Panorama of Jiangou Museum

涧沟陈展馆采用园林景观和建筑空间一体化的设计理念，成为涧沟文化的空间载体。

With the design concept of integrating garden and architectural space, Jiangou Museum has become the spatial carrier of the Jiangou culture.

图 3-051 涧沟陈展馆
Figure 3-051 Jiangou Museum

整个建筑屋顶通过屋顶绿化最大限度使建筑消隐于场地，与自然融为一体。

Through roof greening, the roof of the entire building can be “hidden” and integrated with nature.

图 3-052 涧沟陈展馆下沉广场
Figure 3-052 Sunken plaza at Jiangou Museum

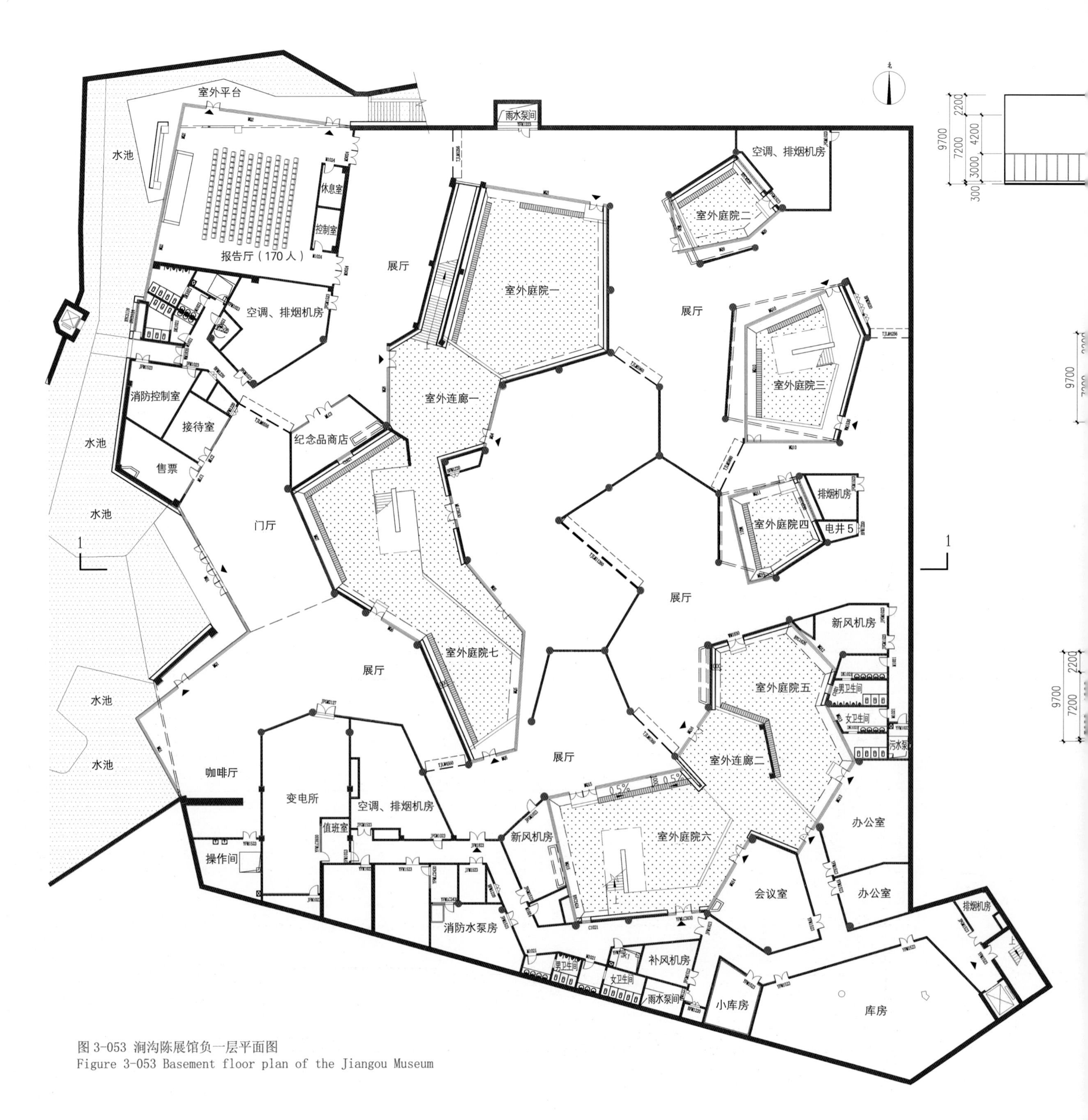

图 3-053 涧沟陈展馆负一层平面图
Figure 3-053 Basement floor plan of the Jiangou Museum

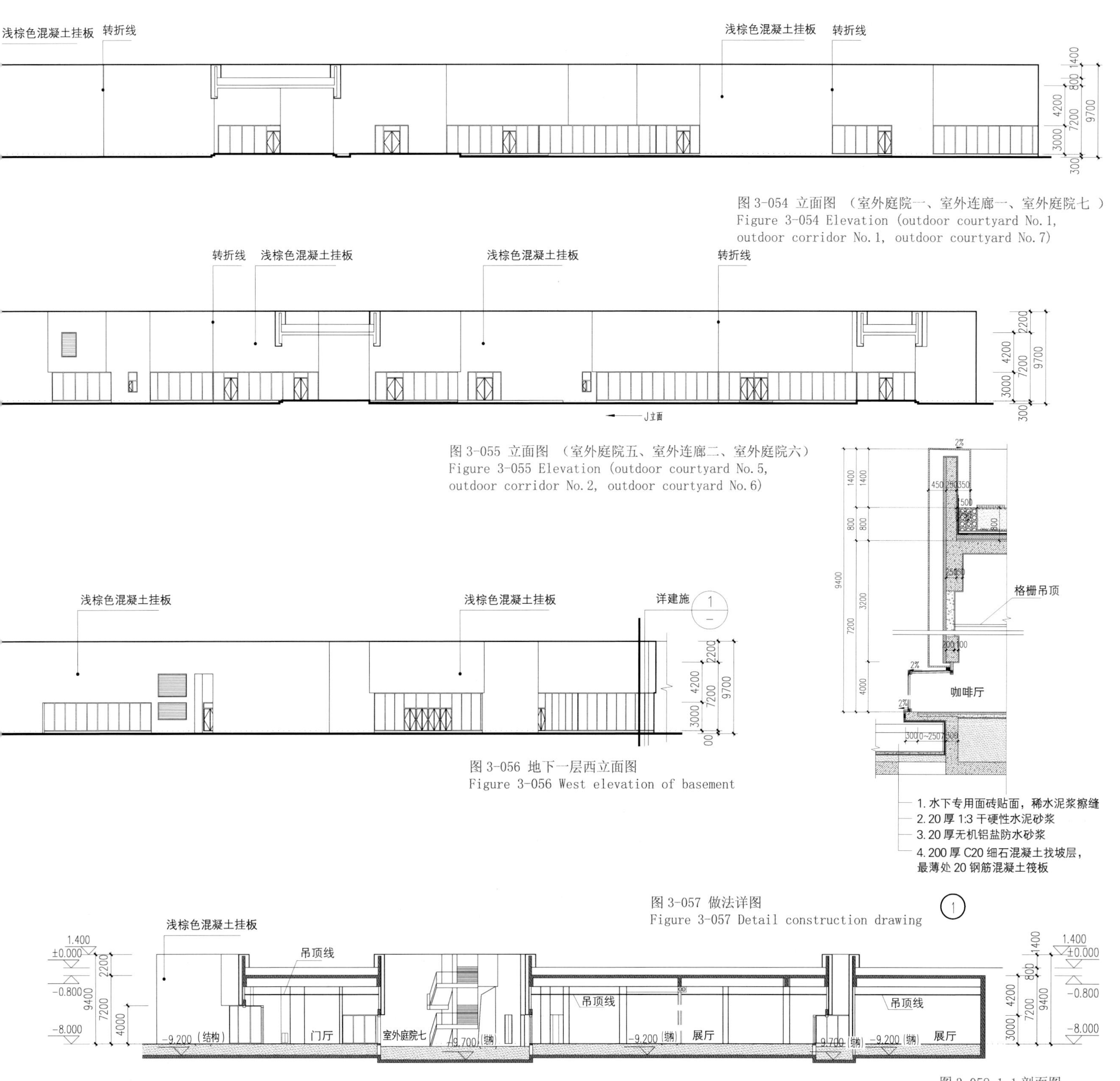

图 3-054 立面图 （室外庭院一、室外连廊一、室外庭院七 ）
Figure 3-054 Elevation (outdoor courtyard No.1, outdoor corridor No.1, outdoor courtyard No.7)

图 3-055 立面图 （室外庭院五、室外连廊二、室外庭院六）
Figure 3-055 Elevation (outdoor courtyard No.5, outdoor corridor No.2, outdoor courtyard No.6)

图 3-056 地下一层西立面图
Figure 3-056 West elevation of basement

图 3-057 做法详图
Figure 3-057 Detail construction drawing

图 3-058 1-1 剖面图
Figure 3-058 Section 1-1

由于建筑紧邻涧沟村历史文化村，故在建筑创作上采取下沉的手法呼应历史遗迹，同时整个建筑下沉以后，建筑体量消逝在大地中，与整个景观体系融为一体。展馆在流线功能设计上采取连续的线性参观路线组织游览，在游览的过程中穿插一系列不同大小和尺度的下沉庭院，这些庭院既保证了室内空间的采光和通风，又作为展览空间的有效疏散，同时还为原本线性的参观流线提供可以穿越的捷径。

Since the building is close to the Jiangou village, sinking techniques were adopted in the architecture of the hall to echo the relics. Meanwhile, the sinking of the entire building makes it "disappear" into the ground and better integrate with the entire landscape. A continuous linear route is adopted for the circulation design. A series of sunken courtyards of different size are interspersed along the circulation process. These sunken courtyards not only ensure the lighting and ventilation of the indoor space, but also serve as additional exhibition space, while providing a shortcut to the originally linear visiting flow.

图 3-059 涧沟陈展馆花田台地
Figure 3-059 Terraced flower field at Jiangou Museum

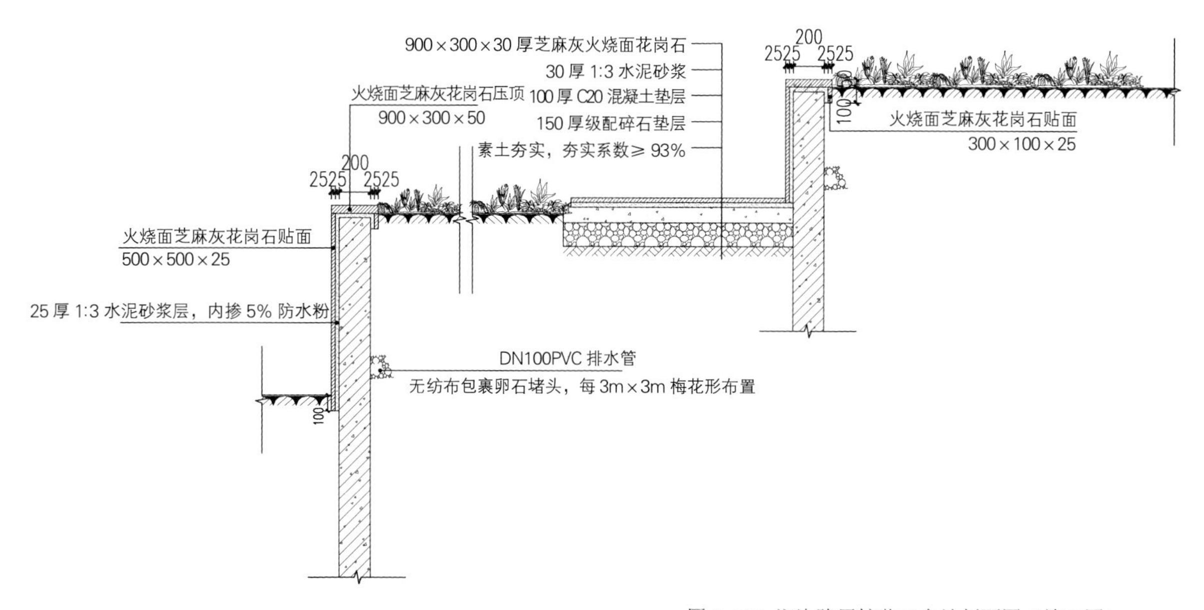

图 3-060 涧沟陈展馆花田台地剖面图（施工图）
Figure 3-060 Section of the terraced flower field at Jiangou Museum (construction drawing)

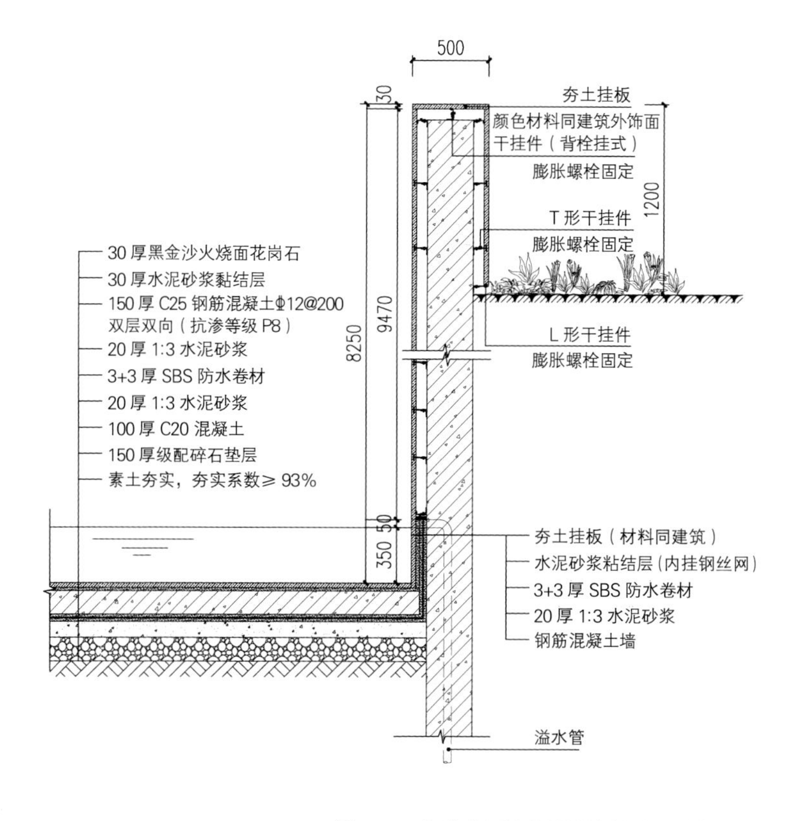

图 3-061 涧沟陈展馆水景景墙剖面图（施工图）
Figure 3-061 Section of the water wall at Jiangou Museum (construction drawing)

3.2.4 景点一 浮光揽月

FLOATING MOON LAKE

浮光揽月场地原为水渣场废弃地，遗留的大面积含有超量的铬、氮和磷的废弃矿渣占据了河道原址，阻隔地表水系，严重污染水质。

面对矿渣废墟，设计以生态修复为核心，在尊重自然的过程中重新梳理水系，构建完善的海绵系统，在安全健康的生态基底上沿湖面组织工业遗风、青山画卷、山水邯郸、赵都新韵等景点，重唤废弃地的勃勃生机。

The Floating Moon Lake site was formerly a waste slag field. A large area of waste slag containing excessive amounts of chromium, nitrogen and phosphorus used to occupy the riverway, blocking the surface water system and seriously polluting the water quality.

Facing the slag ruins, the design centers on ecological restoration, reorganizes the water system while respecting nature. By constructing a complete sponge system, the design creates scenic spots including Industrial Relics, Flowering Terraces, Handan Pavilion, and the New Handan Pavilion along the lake surface on a safe and healthy ecological base, recalling the vitality of the abandoned land.

图 3-062 浮光揽月现状实景

Figure 3-062 Photo showing existing condition of Floating Moon Lake

图 3-063 浮光揽月鸟瞰图（主入口视角）
Figure 3-063 Bird eye view of the Floating Moon Lake looking from the main entrance (

设计特色

DESIGN FEATURES

（1）连通水系打造海绵基底

邯郸母亲河——沁河的连通与纯净，关系着整个城市水生态的安全。利用场地残存矿坑洼地挖深垫浅打造开阔水面，形成园区最大的人工湖——揽月湖，重新连接沁河上下游，恢复水体生态过程的完整性和系统性。

湖中悬浮若干利用本地毛石块堆砌的树岛，湖边种植荷花、黄菖蒲等湿地植物，利用植物的蒸腾与吸附作用净化水体，自然做功恢复水体生态系统的健康。水体与多样的植物同时为野生动物营造出丰富的自然生境，增加生物多样性。

通过湖边的雨水收集带和雨水花园实现雨水的收集利用。雨水收集带和雨水花园一般为浅凹绿地的形式，下雨时可吸收蓄积周边雨水，通过植物及砂土等透水材料使雨水得到过滤净化，并使之回渗土壤中，在需要时又可将蓄存的水释放并加以利用，形成良好的循环利用模式。

雨水花园中多余的雨水通过竖向设计汇入揽月湖中，湖体与雨水收集带、雨水花园作为园区水生态基础设施，共同构成海绵城市系统，使城市能够像海绵一样在面对环境变化和雨水灾害方面具有良好的弹性和适应性。

（1）Connecting the Water System to Create a Sponge Base

The connection and purity of Qin River is a key to the safety of the entire city's water ecology. By using the remaining pits in the site, an open water surface is created that is the largest artificial lake in the park—The Floating Moon Lake, in order to reconnect the upstream and downstream of Qin River, and restore the integrity and systematicness of the ecological process of the water body.

Several tree islands built with local rubble stones are floating in the lake. Wetland plants such as lotus and yellow calamus are planted by the lake. The transpiration and adsorption of the plants are used to purify the water body, letting nature restore the health of the water ecosystem. The water body and diverse plants create a rich natural habitat for wild animals while increasing biodiversity.

Rainwater collection and utilization are realized through the rainwater collection belt and rainwater garden, which are generally in the form of shallow concave green spaces. When it rains, they can absorb and accumulate surrounding rainwater. The rainwater then will be filtered and purified by permeable materials such as plants and sand. The rainwater can be returned to the soil and, when needed, released and reused again.

Through grading, excess rainwater in the rainwater garden flows into the Floating Moon Lake. The lake body, the rainwater collection belt and the rainwater garden are used as the water ecological infrastructure of the park, which together constitute a sponge city system, enabling the city to face environmental changes and rain disasters with higher resilience and flexibility.

图 3-064 荷叶挺立在水中连成一片
Figure 3-064 Lotus leaves floating in the water

图 3-065 湖中悬浮圆形树岛
Figure 3-065 Round tree islands floating in the lake

（2）湖光流影映衬景区核心

基于自然生态过程打造的揽月湖成为核心游览区的中心，水面如镜。湖畔赵都新韵、山水邯郸、工业遗风等特色核心景点环绕，倒影绰绰映水中，静时水光浮翠，动时波光粼粼，打造了一个独具特色的园博园标志性景观群落。

（2）Main Attractions Reflected in the Lake

The Floating Moon Lake built based on the natural ecological process has become the center of the core tourism area. Main attractions such as New Handan Pavilion, Handan Pavilion and Industrial Heritage Pavilion, are reflected in the water, shining in static and sparkling in movement, creating a unique iconic landscape community of the park.

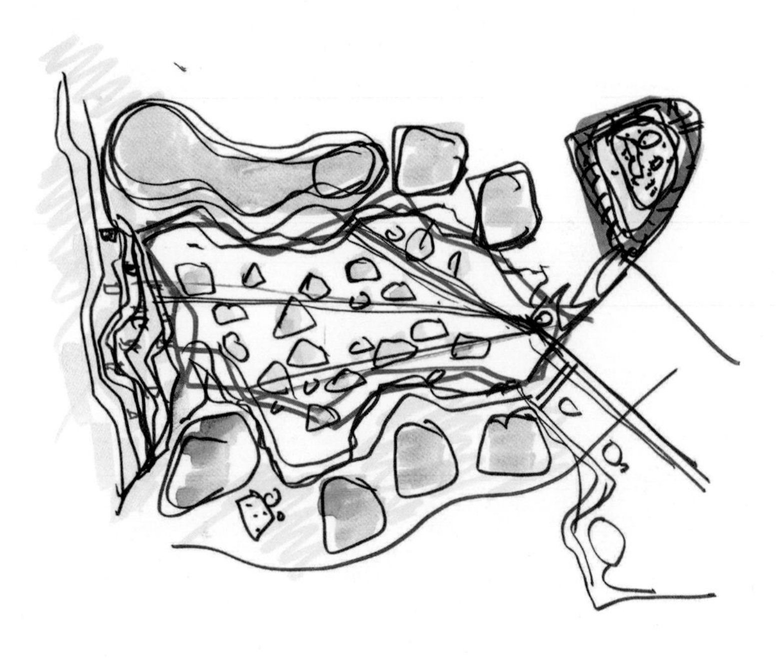

图 3-066 首席设计师俞孔坚浮光揽月节点草图
Figure 3-066 Sketch drawing of Floating Moon Lake by chief designer Yu Kongjian

图 3-067 浮光揽月全景
Figure 3-067 Panorama of Floating Moon Lake

环湖步道提供不同的观景角度，蜿蜒延伸的主路和多条支路反复交织形成灵活便捷的路网，创造多次相逢与分离的机遇，步道间散布的树岛营造出宜人的微气候。

The walking trail around the lake provides different viewing angles. The winding main path and multiple branch paths are repeatedly interwoven to form a flexible and convenient trail network, creating opportunities for multiple encounters and separations. The tree islands scattered among the trails create pleasant micro climates.

图 3-068 浮光揽月剖面图
Figure 3-068 Section of the Floating Moon Lake

图 3-069 浮光揽月多元慢行系统
Figure 3-069 Slow-traffic system at the Floating Moon Lake

图 3-070 休憩景观盒为游客提供了半私密的空间
Figure 3-070 The landscape lookout provides a semi-public space for visitors

图 3-071 夕阳下环湖散步的游客
Figure 3-071 Visitors taking walks around the lake in the sunset

（3）水上灯光丰富夜间体验

夜晚的揽月湖水面平静沉寂，成为水上灯光秀的背景。华灯初上之时湖中将呈现纷繁的水上灯光表演。表演结束，湖面即刻恢复月下天镜之状，体现园博园夜色下的静谧之美。

（3）Lighting in Water Enriches Night Experience

The calm water of the Floating Moon Lake at night becomes the background of the water light show. When the lanterns light up, there will be a variety of water light shows in the lake. At the end of the performance, the lake immediately regained its appearance as a mirror under the moon, reverting the park's tranquil beauty at night.

图 3-072 灯光为景，湖面为幕
Figure 3-072 Night light relfects on the lake

图 3-073 会后成为市民广场
Figure 3-073 Turning into a civic square after Expo

图 3-074 滨水栈道
Figure 3-074 Waterfront boardwalks

图 3-075 景观盒
Figure 3-075 Lookout pergola

图 3-076 水秀表演
Figure 3-076 Water show

3.2.5 景点二 赵都新韵

THE NEW HANDAN PAVILION

赵都新韵位于揽月湖湖畔，是本届园博会主办方——邯郸市的展示园区，是一座“开放式的园林博物馆”，集中展示邯郸特色地域文化及造园新技术与新材料。设计以“多宝格”作为创作灵感，邯郸当地特色文化作为格子上的宝物呈现，游客在“宝格”中穿梭，移步异景，领略邯郸深厚的文化底蕴和丰富的历史内涵。

赵都新韵内的展园可分为核心展示区和游客服务区两部分。核心展示区以一条双层时空走廊作为纽带，将两个出入口、休息区、6 个特色展园、竹林休憩区及全息影像馆串联起来。游客服务区位于展园南侧，结合三处游客服务建筑及其前面的景观休憩广场，提供售卖、餐饮休憩、纪念品商店及厕所等功能，为游客提供舒适的游览、休憩体验。

Located by the Floating Moon Lake, the New Handan Pavilion is the exhibition garden of Handan City. It is an “open garden museum” that focuses on exhibiting Handan’s characteristic regional culture, new techniques and new materials for gardening. The design is inspired by “gem grids”, where local characteristic culture of Handan is presented as a gem in a grid. Visitors roam through the “gem grids” of different scenes, and appreciate Handan’s profound cultural heritage and rich historical connotation.

The exhibition garden in the New Handan Pavilion can be divided into two parts: the core exhibition area and the tourist service area. The former is linked by a double-level corridor, connecting two entrances, the rest areas, six theme exhibition gardens, the Bamboo Pergola and Hologram Pavilion. The tourist service area is located on the south side of the exhibition garden. It combines the three tourist service buildings and the landscape recreation square in front of it. It provides functions including retails, dining and resting, souvenir shops and toilets.

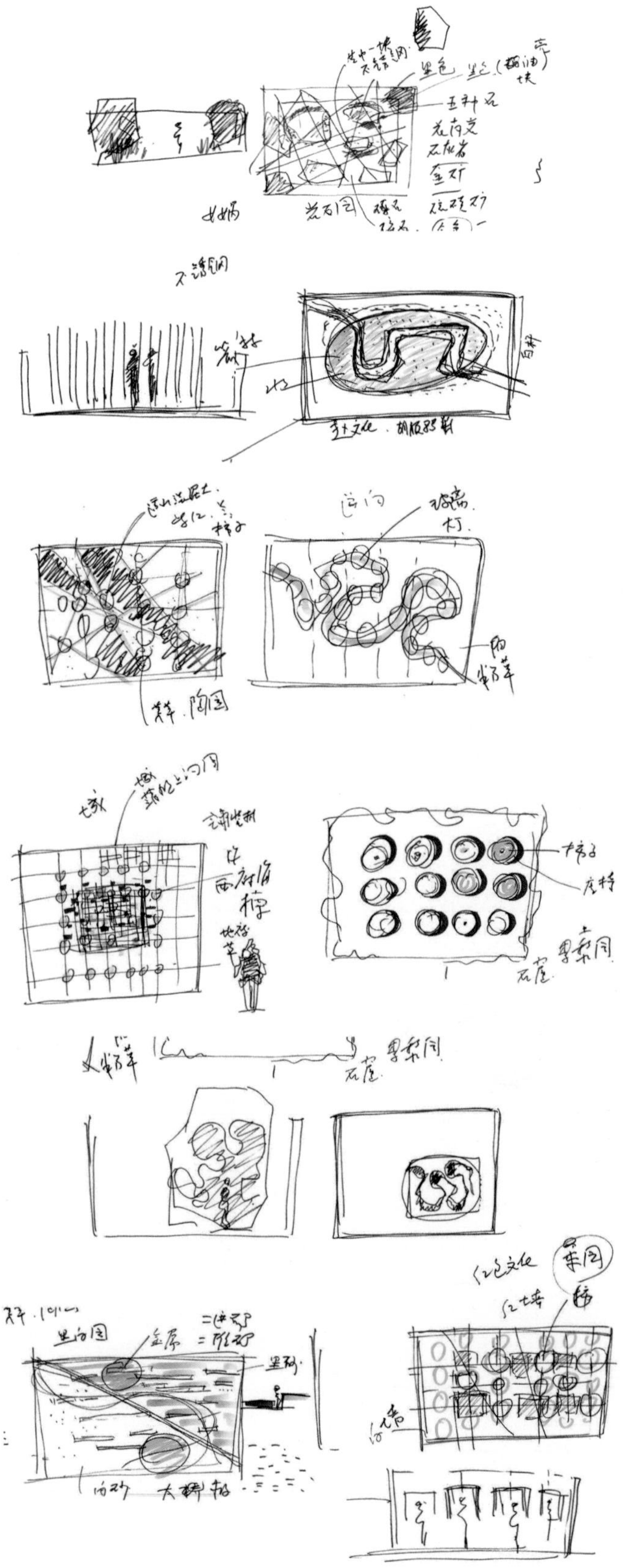

图 3-077 首席设计师俞孔坚赵都新韵节点草图
Figure 3-077 Sketch drawings of the New Handan Pavilion by chief designer Yu Kongjian

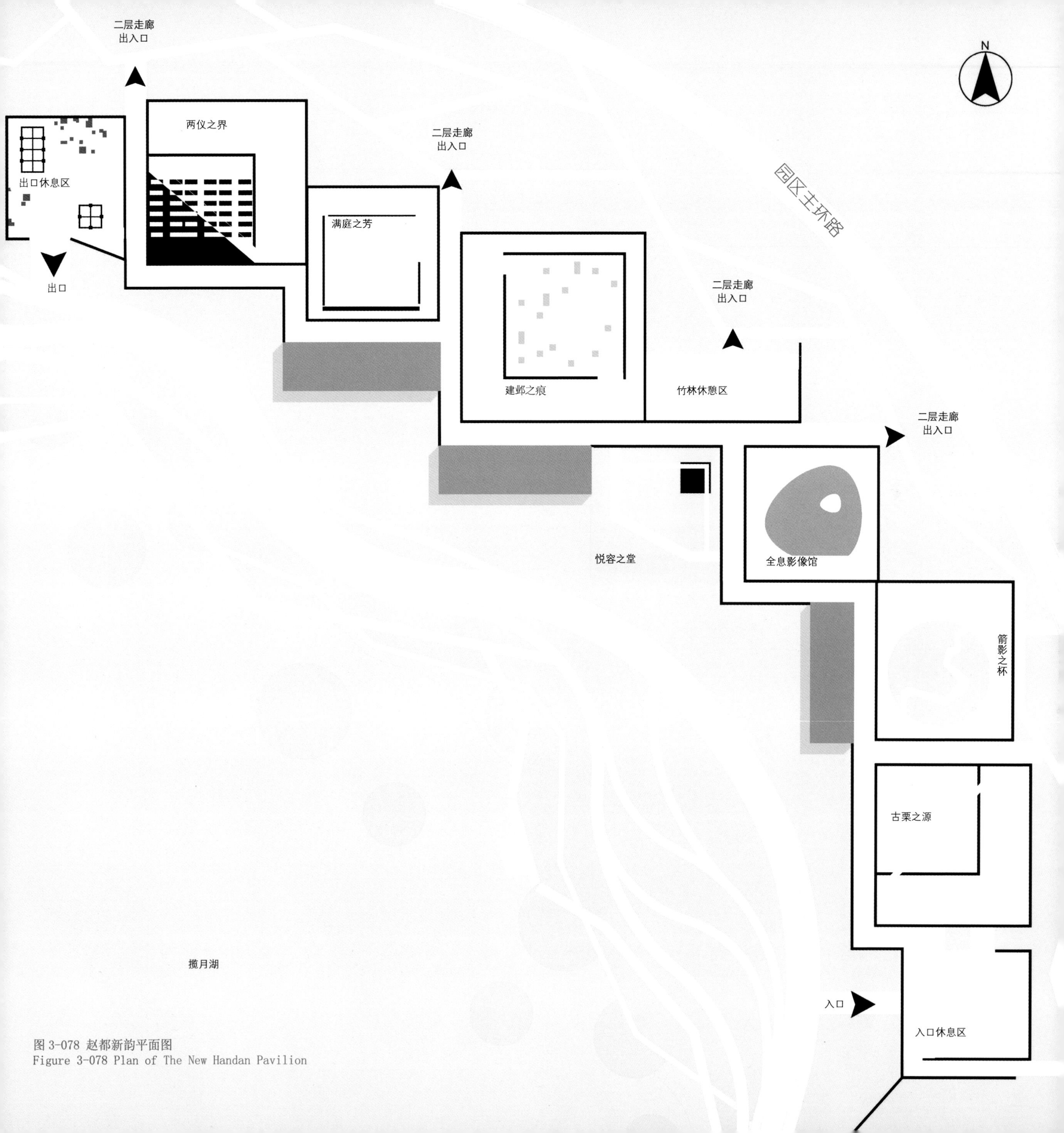

二层走廊
出入口

两仪之界

出口休息区

出口

二层走廊
出入口

满庭之芳

园区主环路

建邺之痕

二层走廊
出入口

竹林休憩区

二层走廊
出入口

悦容之堂

全息影像馆

箭影之杯

古栗之源

揽月湖

入口

入口休息区

图 3-078 赵都新韵平面图
Figure 3-078 Plan of The New Handan Pavilion

图 3-079 赵都新韵鸟瞰图（出口方向）
Figure 3-079 Bird-eye view of the New Handan Pavilion (viewing from exit)

时空走廊依据周边场地标高设计为双层，上层与外部电瓶车道相连，下层与浮光揽月相通，上下两层空间通过不同形式的台阶联系起来。游廊由不同厚度的青砖交替拼接而成，呈现凹凸感，局部间隔呈门状造型，时而呈现框景效果，时而作为入口存在。

The Time Corridor is designed as a double-level according to the elevation of the surrounding site. The upper level is connected with the external cart lane, while the lower level connects with the Floating Moon Lake, and the two levels are connected by different forms of steps. The corridor is spliced by black bricks of different thicknesses, presenting a sense of unevenness, with the intervals in a door-like shape, making the corridor sometimes present a framed effect, and sometimes function as an entrance.

图 3-080 时空走廊（下层）
Figure 3-080 The Time Corridor (ground level)

图 3-081 时空走廊（上层）
Figure 3-081 The Time Corridor (upper level)

图 3-082 混凝土块拼花墙（左）与青砖拼花墙（右）
Figure 3-082 Concrete motif wall (left) and black porcelain motif wall (right)

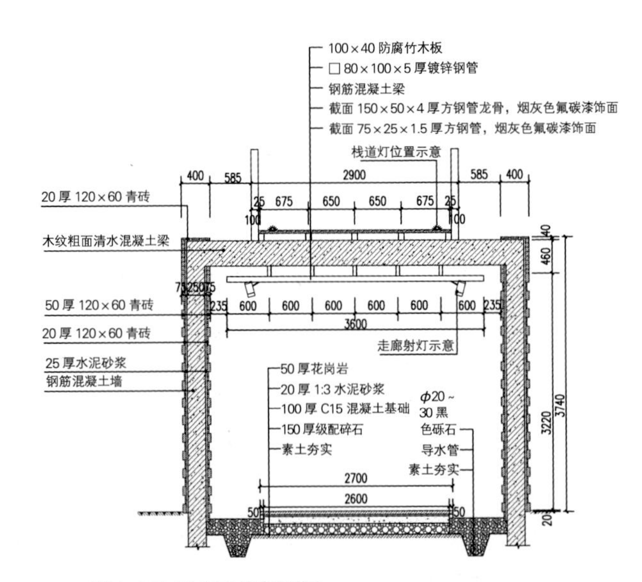

图 3-083 时空走廊剖面图一
Figure 3-083 Construction drawing of the Time Corridor

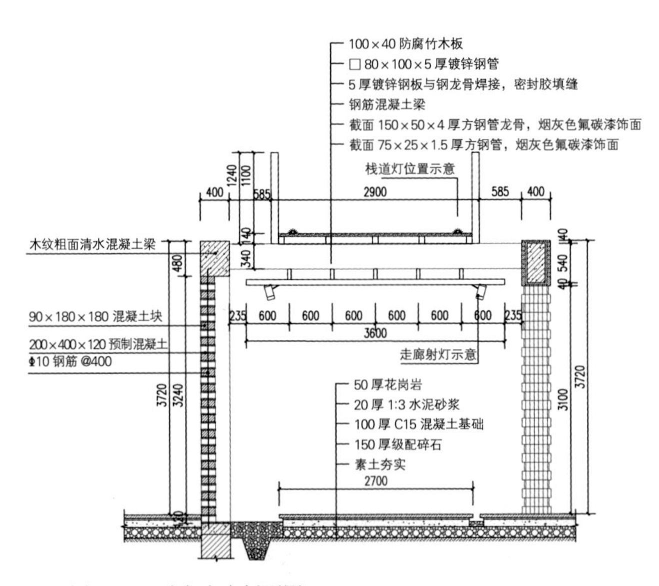

图 3-084 时空走廊剖面图二
Figure 3-084 Construction drawing of the Time Corridor

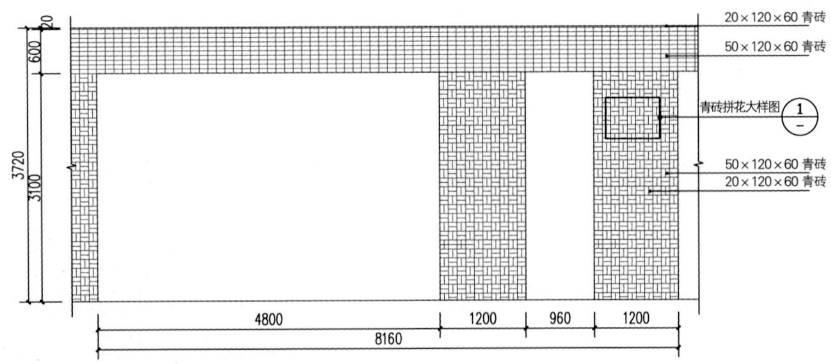

图 3-085 文化墙标准段一立面图
Figure 3-085 Typical elevation of the cultural wall

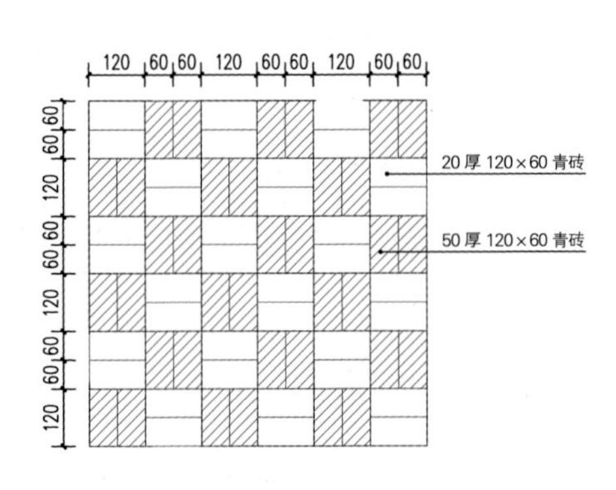

图 3-086 青砖拼花大样图
Figure 3-086 Detail drawing of black brick tiles

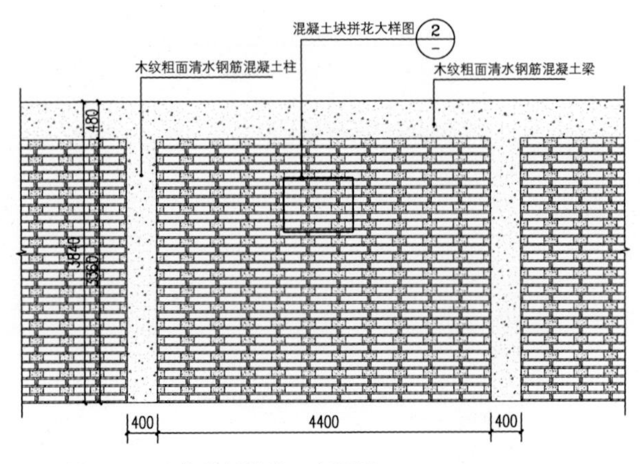

图 3-087 文化墙标准段二立面图
Figure 3-087 Typical elevation of the cultural wall

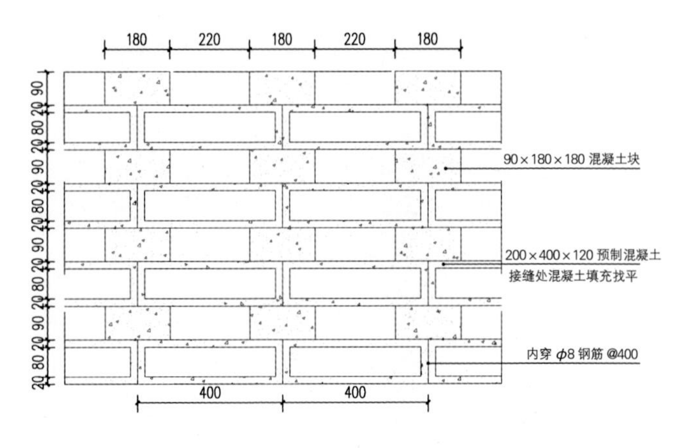

图 3-088 混凝土块拼花大样图
Figure 3-088 Detail drawing of concrete tiles

展园特色详解
GARDEN DETAILS

（1）入口展园

入口展园作为赵都新韵的序厅，占地面积 475m^2。空间布局借鉴了欲扬先抑的造园手法，入口处设置一处绿荫廊架，由 12 根钢筋混凝土柱和屋顶绿化组合而成，整体呈现生态简洁的效果。

（1）Entrance Exhibition Garden

As the preface of the New Handan Pavilion, the entrance exhibition garden covers an area of 475 square meters. Its spatial layout draws on the gardening technique of “promote by suppressing”. A pergola is set at the entrance, which is composed of 12 reinforced concrete pillars and roof greening.

内部由蓝色铝板墙半围合成方形空间，墙面和地面散布各色景石，沿墙体布置雾喷，营造出远古时期女娲五色石补天仙境般的氛围，是对邯郸女娲文化的一种体现。

The interior is half-enclosed by a blue aluminum plate wall to form a square space. The walls and the floor are scattered with various scenery stones. Fog sprays are arranged along the walls to resemble the atmosphere of Nuwa mending the sky with the five-color stones—creating a manifestation of Handan’s Nuwa culture.

图 3-089 入口展园效果图
Figure 3-089 Entry garden rendering

（2）古粟之源

磁山文化起源于河北省邯郸市武安磁山，是北方旱作农业的“谷子种植”发源地，具体可以追溯到距今8000多年的新石器时期。

（2）The Garden of Millet

The Cishan culture originated in Cishan, Wu'an, Handan City. It is the birthplace of the north China's dry farming "millet planting". It can be traced back to the Neolithic Age more than 8,000 years ago.

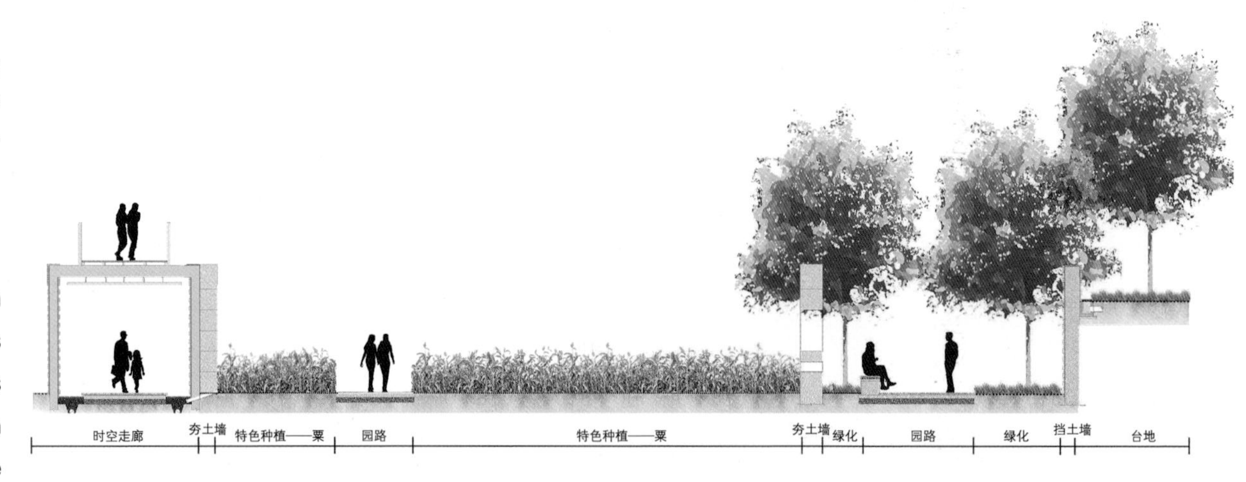

图 3-090 古粟之源剖面图
Figure 3-090 Section of Garden of Millet

图 3-091 古粟之源与入口展园鸟瞰图
Figure 3-091 Bird eye view of Garden and of Millet and Entrance Pergola

古粟之源占地面积 900m^2，分为展览区和休憩区两个部分。展览区位于西北侧，以磁山文化为设计依据，由夯土墙围合为方形。

The Garden of Millet covers an area of 900 square meters and is divided into an exhibition area and a rest area. The exhibition area is located on the northwest side. It is designed based on the Cishan culture and is enclosed by rammed earth walls into a square shape.

墙体颜色以土黄色系为主，内嵌新石器时期的破碎陶瓷片等，利用磁山文化的装饰元素，展现 7000 多年前人类的生活方式。地面种植粟米，一条混凝土栈道穿行而过连接古今，行走其中，仿佛置身于古人劳作场景，立体地展示邯郸农业文化传统的魅力。

The main color of the wall is earthy yellow, and elements that reflect the Cishan culture, such as broken ceramic pieces from the Neolithic period, are embedded to demonstrate the way of life more than 7,000 years ago. Millet is planted on the ground, and a concrete boardwalk passes through to connect the past and the present. Walking through it, the visitor feels like walking in the ancient farming scene. The charm of Handan's agricultural culture is displayed in a three-dimensional manner.

图 3-092 古粟之源
Figure 3-092 The Garden of Millet

（3）箭影之林

战国初期，赵国为逐鹿中原而迁都邯郸，以“胡服骑射”为代表的军事改革展示了赵文化开放、进取、包容的内涵。

（3）The Garden of Arrows

At the beginning of the Warring States Period, State of Zhao moved its capital to Handan in order to compete in the Central Plains. By putting forward the military reform represented by “Riding and Shooting in Hu Clothing”, Zhao demonstrated the openness, enterprising spirit, and tolerance in its culture.

箭影之林占地面积 650m^2，运用了天圆地方的设计理念，整体布局为方形场地内嵌圆形水池。

The Garden of Arrow covers an area of 650 square meters. Its square site with a circular pool embedded in it represents the concept of “round sky and square earth”.

图 3-093 箭影之林鸟瞰图
Figure 3-093 Bird eye view of the Garden of Arrows

展园以逐鹿中原时的战争场景为启发，水池中树立不锈钢 “箭林”，展示金戈铁马的战场风貌，柱阵在夜晚呈现五彩变换的灯光效果，为游客带来不同的视觉体验。

The exhibition garden is inspired by the war scenes. The stainless steel “arrow forest” in the pool shows the style and features of the battlefield. The array of arrows presents colorful lighting effects at night, bringing visitors a different visual experience.

图 3-094 不锈钢“箭林”
Figure 3-094 “Arrow forest” made of stainless steel

水池边缘竖立一组印有“胡服骑射”图案的冲孔钢板景墙，利用冲孔大小变化构成的胡服骑射历史故事图案，展现赵国的改革创新精神。

A set of perforated steel plate scenery walls printed with the pattern of “Riding and Shooting in Hu’s Clothing” are erected on the edge of the pool. The historical stories composed of changes in the size of the perforation is used to show the spirit of innovation of the State of Zhao.

图 3-095 “胡服骑射”景墙
Figure 3-095 A feature wall exhibits the story of “Riding and Shooting in Hu’s Clothing”

（4）全息影像馆

全息影像馆灵感来自于邺陶古窑址传统圆形窑，占地面积 650m²，是一处以影像的方式展示邯郸文化的休憩场所。

影像馆采用红砖砌筑，类似倒扣碗状，拱顶中央顶部开圆形洞口，增加室内光影的变化。馆内为下沉式空间，与高科技环幕投影相结合，为游客打造一个沉浸式体验空间，用现代科技手法传递邯郸文化的历史变迁。

（4）Hologram Pavilion

Hologram Pavilion is inspired by the traditional round kiln at the Yetao Ancient Kiln Site. It covers an area of 650 square meters and is a resting place that displays Handan culture in the form of images.

The pavilion, in the shape of an inverted bowl, is built with red bricks. The circular opening on the top of the arch increases the variation of indoor light and shadow. The inside of the pavilion is a sunken space, combined with high-tech circular screen projection, creating an immersive experience space for visitors. The historical changes of Handan culture are demonstrated with modern technology.

图 3-096 全息影像馆
Figure 3-096 Hologram Pavilion

图 3-097 悦容之堂和全息影像馆鸟瞰图
Figure 3-097 Bird-eye view of the Garden of Smiling Buddha and the Hologram Pavilion

（5）悦容之堂

位于邯郸市的响堂山石窟是北齐佛教造像艺术的代表，共有大小洞窟 16 座，造像 4300 多尊，是北齐佛教雕刻艺术的宝藏。

（5）Garden of Smiling Buddha

The Xiangtangshan Grottoes in Handan City represent the Buddhist sculpture art of the Northern Qi Dynasty. There are 16 caves of all sizes and more than 4,300 sculptures.

悦容之堂以北齐石窟为主题，占地面积 680m^2，展园内的两条栈道交叉漂浮于水面上，双侧有喷雾设施，提升游客的观园体验，池面种植睡莲，营造“浮香绕岸，圆影覆池”的灵动空间。

The Garden of Smiling Buddha takes the Northern Qi Grottoes as the theme. It covers an area of 680 square meters. The two boardwalks float diagonally on the water surface. Spray facilities are put on both sides to enrich visitors' experience. Water lilies are planted on the pool surface to create a smart space.

图 3-098 悦容之堂
Figure 3-098 The Garden of Smiling Buddha

园内高塔是赵都新韵的制高点。塔高12m，立面肌理灵感来自邯郸出土佛教造像“褒衣博带”的特征，内部展示以黄铜材质复刻邯郸博物馆出土文物“邯郸微笑”。

The high tower in the park is the commanding height of New Handan Pavilion. The tower is 12 meters high, and the facade texture is inspired by the characteristics of the Buddhist statue “Baoyi Bodai” unearthed in Handan. The internal exhibition uses brass to reproduce the cultural relic “Smiling Buddha” of the Handan Museum.

图 3-099 邯郸微笑
Figure 3-099 Smiling Buddha

图 3-100 竹林休憩区
Figure 3-100 Bamboo Pergola

（6）竹林休憩区

竹林休憩区是赵都新韵上下两层交通的连接处之一，场地整体呈长方形，占地 490m^2, 利用台地的方式来解决场地的高差及交通动线问题。台地共 4 层，由步道相连，每一层有两级台地，台地内种满了竹子，每一层的下级台地表面铺设防腐木坐凳。竹林下静谧、阴凉，是园内主要的游客休息场所。

（6）Bamboo Pergola

Bamboo Pergola is one of the traffic connections between the upper and lower levels of the New Handan Pavilion. The site is rectangular and covers an area of 490 square meters. Terraces are used to treat the height difference and traffic flow of the site. There are four levels of terraces, connected by walkways. Each level has two layers of terraces, which are planted with bamboo. The surface of the lower layer of each level is equipped with anticorrosive wood benches. The quiet and shady area under the bamboo forest provides a resting place for visitors.

（7）建邺之痕

邺城在我国城市建筑史上占有重要地位，堪称中国城市建设的典范，本园就是对这一建筑成就的致敬和传承。建邺之痕呈方形，占地面积 1120m^2，是赵都新韵最大的展园。

展园设计从邺城城郭布局出发，提取邺城“匠人营国，方九里，旁三门。国中九经九纬，经涂九轨，左祖右社，面朝后市，市朝一夫”的营城理念。形制简化为阡陌纵横的马赛克图案肌理，斑驳的青色墙体则是对古邺城城墙的映像，使人在尺寸之间体会中国古人的建城智慧。

（7）The Garden of Relic City

Being called a model of China's urban construction, Ye City lies on an important position in the history of China's urban architecture. Being a tribute to and inheritance of Ye City's architectural achievements, Garden of Relic City is square in shape and covers an area of 1,120 square meters. It is the largest exhibition garden of the New Handan Pavilion.

The design of the garden mirrors the layout of the Ye City. It complies with Ye City's allusion "The architect builds the city as a nine-li-square with three gates on each side. The city has nine streets transverse and nine longitudinal, with each street accommodating nine carts in parallel. Temples sit on the left side of the streets, and altars on the right. Royal courts sit in front and markets on the back, both the courts and markets are 100-step square each". The shape of the ancient city is simplified into mosaic patterns in vertical and horizontal rows, and the mottled cyan wall is a reflection of the ancient city wall of Ye City, allowing visitors to experience the wisdom of ancient Chinese city in the sense of scale.

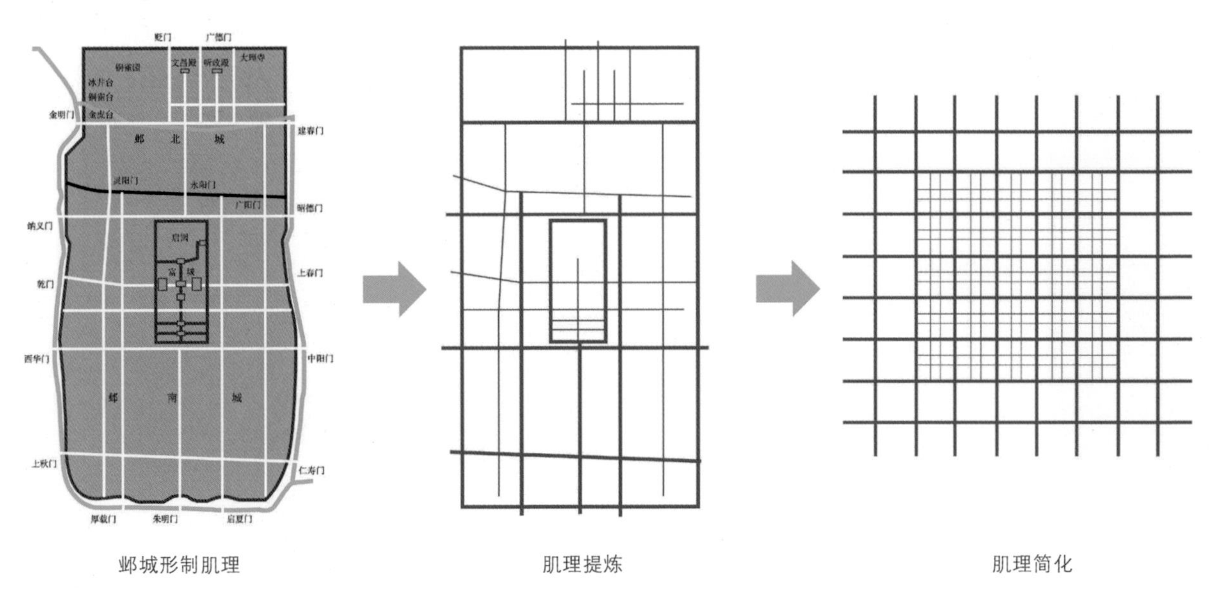

图 3-101 建邺之痕设计思路
Figure 3-101 Design process of the Garden of Relic City

图 3-102 建邺之痕鸟瞰图

Figure 3-102 Bird eye view of the Garden of Relic City

中心区由旧的青砖、青瓦拼接而成的残墙围合而成，展示邺城曾为六朝古都的辉煌历史。

The central area is enclosed by the remnant walls made up of old cyan bricks and tiles, showcasing the glorious history of Ye City as the ancient capital of the Six Dynasties.

图 3-103 青砖、青瓦拼接的残墙
Figure 3-103 Broken walls composed by black bricks and black porcelain tiles

地面铺砌马赛克种植槽，槽内侧沿线放置雾喷装置，绿地旁设置黄色玻璃钢坐凳。

Mosaic planting pools are paved on the ground, spraying devices are placed along the inside of the pools, and yellow glass fiber reinforced plastic benches are set next to the planting pools.

图 3-104 建邺之痕中心区
Figure 3-104 Central area of the Garden of Relic City

（8）满芳之庭

在邯郸出土的磁州窑的诸多品种中，尤以白地黑花（铁锈花）、刻划花、窑变黑釉最为著名。其中，具有水墨画风的白地黑绘装饰艺术吸收了传统的水墨画和书法艺术的技法，开启了中国瓷器彩绘装饰的先河。

满芳之庭占地面积630m^2，中心广场以磁州窑白底黑绘装饰手法为灵感设计地面铺装。因磁州当地的瓷土含有少量铁粉，高温烧制后呈灰白色或灰黄色，所以广场铺装选用瓷器原始状态胚体色——土黄色为底色，绘制黑色植物纹样肌理，给人以强烈的视觉冲击。展园内部结合纹样肌理设置四组白色圆形座椅，使置身其中的游客在这原始的瓷器工艺中体会古代匠人的精湛工艺。

（8）The Garden of Ceramics

Among the many varieties of the Cizhou kiln unearthed in Handan, the most famous ones are the black floral design over white ground (rust flower), the carved flower, and the blackened kiln glaze. Among them, the white ground with black floral decorative art features the traditional ink painting and calligraphy art techniques, which opened the precedent of Chinese porcelain painting decoration.

The Garden of Ceramics covers an area of 630 square meters, with the paving of its central square inspired by the black floral painting over white ground decoration technique of the Cizhou kiln. The local porcelain clay in Cizhou contains a small amount of iron powder, it becomes grayish white or grayish yellow after high temperature firing. Therefore, the original earthy yellow color of the porcelain is used as the background color for the square paving, and the black plant pattern giving visitors a strong visual impact. Four groups of white round seats are set up in combination with the pattern texture within the garden, so that visitors can experience the exquisite craftsmanship of ancient craftsmen in the original porcelain craftsmanship.

图3-105 满芳之庭鸟瞰图
Figure 3-105 Bird eye view of the Garden of Ceramics

北侧的一组瓷片墙，由方钢搭建整体框架，内嵌钢筋串联瓷片，每一组瓷片由白黑两色组成，可随意翻动拼出各种图案，增加园景的趣味性及互动性。

其他三面由“笼盔”特色墙营造出虚实变化的景观。笼盔上的洞可以透光，人们还可以“窥视”园内园外的景色，别有一番乐趣。

The wall build of porcelain tiles on the north side features a frame that is made of square steel, with steel bars embedded to connect the tiles. Each group of the tiles is composed of white and black colors, which can be flipped to create a variety of patterns and increase the interest and interactivity of the garden.

On the other three sides, the “Longkui” wall creates a changeable landscape between the virtual and the real. The holes in the “Longkui” allow light to pass through, and people can “peek” through at the scenery inside and outside the garden.

图 3-106 瓷片墙及“笼盔”墙
Figure 3-106 The porcelain tile wall and the “Longkui” wall

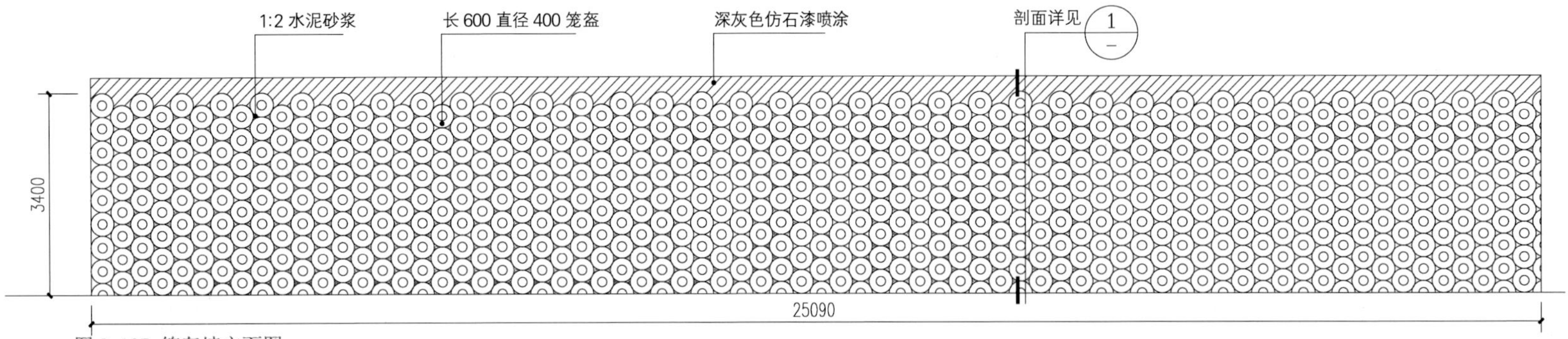

图 3-107 笼盔墙立面图一
Figure 3-107 Elevation of "Longkui" wall (I)

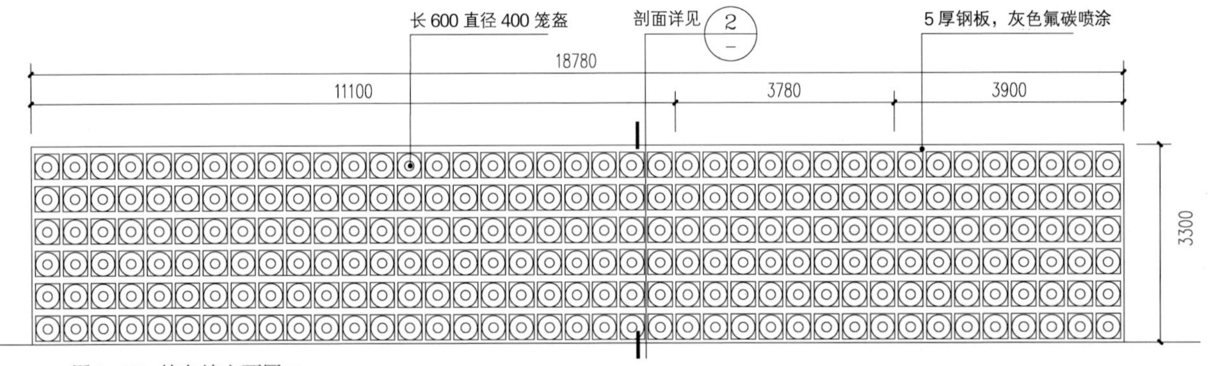

图 3-108 笼盔墙立面图二
Figure 3-108 Elevation of "Longkui" wall (II)

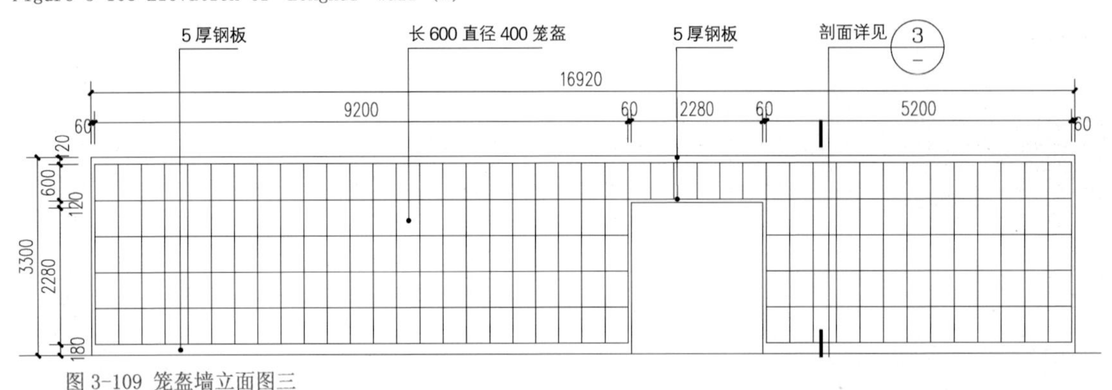

图 3-109 笼盔墙立面图三
Figure 3-109 Elevation of "Longkui" wall (III)

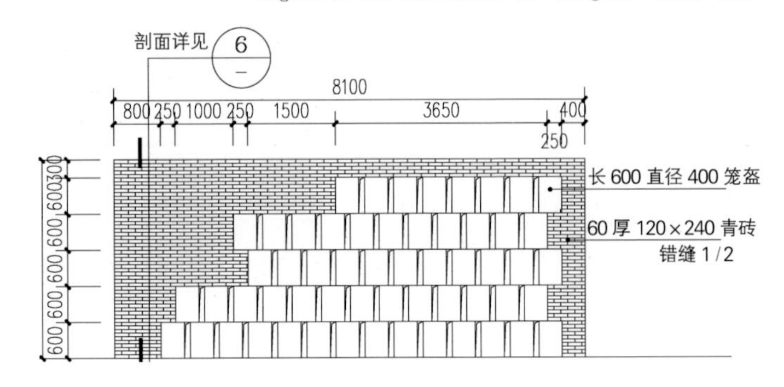

图 3-110 笼盔墙立面图四
Figure 3-110 Elevation of "Longkui" wall (IV)

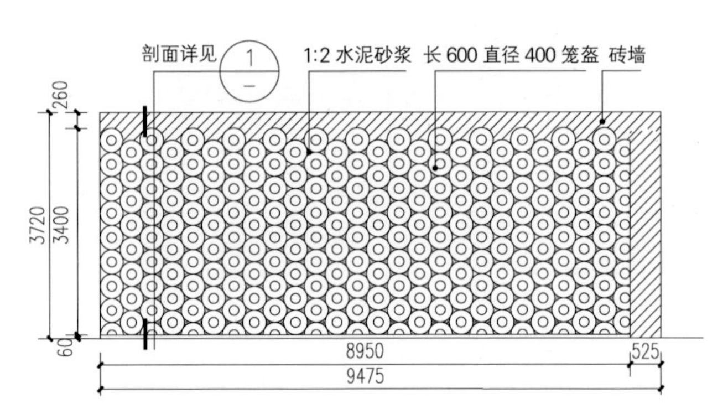

图 3-111 笼盔墙立面图五
Figure 3-111 Elevation of "Longkui" wall (V)

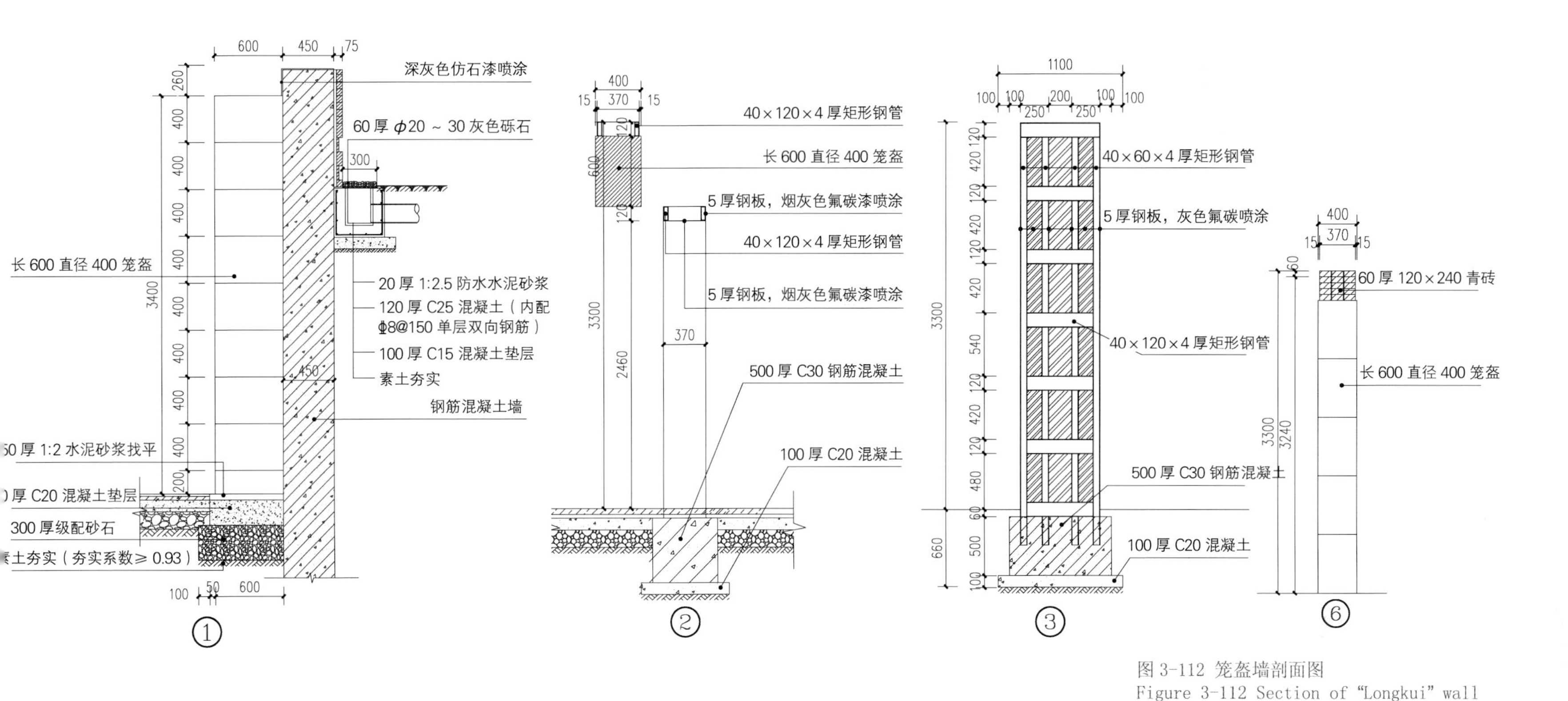

图 3-112 笼盔墙剖面图
Figure 3-112 Section of "Longkui" wall

□ 120 × 120 × 10 方钢管

剖面详见 ④

大样详见 ⑤

2075 3070 3280 3280 3070 2075

16850

图 3-113 瓷片墙立面图
Figure 3-113 Elevation of porcelain wall

25 厚 200 × 100 白色 / 黑色瓷砖

10 厚 φ40 钢片

φ10 钢柱

密封胶

φ10 钢柱

25 厚 200 × 100 白色瓷砖

25 厚 200 × 100 黑色瓷砖

10 厚 ф40 钢片

⑤ 图 3-114 陶瓷装饰隔断节点详图
Figure 3-114 Detail drawing of porcelain joints

□ 120 × 120 × 10 方钢管

ф10 钢柱

25 厚 200 × 100 白色 / 黄色瓷砖

10 厚 φ40 钢片

20 厚瓷砖面层同周围铺装材料

30 厚 1:3 水泥砂浆

100 厚 C15 混凝土

150 厚级配碎石

素土夯实

④ 图 3-115 瓷片墙剖面图
Figure 3-115 Section of porcelain wall

（9）两仪之界

太极拳八大门派之五源于邯郸永年广府，被誉为“中国太极之乡”。太极的理念延伸体现为太极八卦图，“两仪生四象，四象生八卦”是中国古代道家论述万物变化的重要原理。

（9）The Garden of Tai Chi

The fifth of the eight schools of Tai Chi originated in Yongnian Guangfu, Handan, which is known as the “Hometown of Chinese Tai Chi”. The extension of the concept of Tai Chi is embodied in the Eight Diagrams of Tai Chi, “The Two Modes generate the Four Images, and the Four Images generate the Eight Diagrams”, which is one of the important principles of ancient Chinese Taoists discussing the changes of all things.

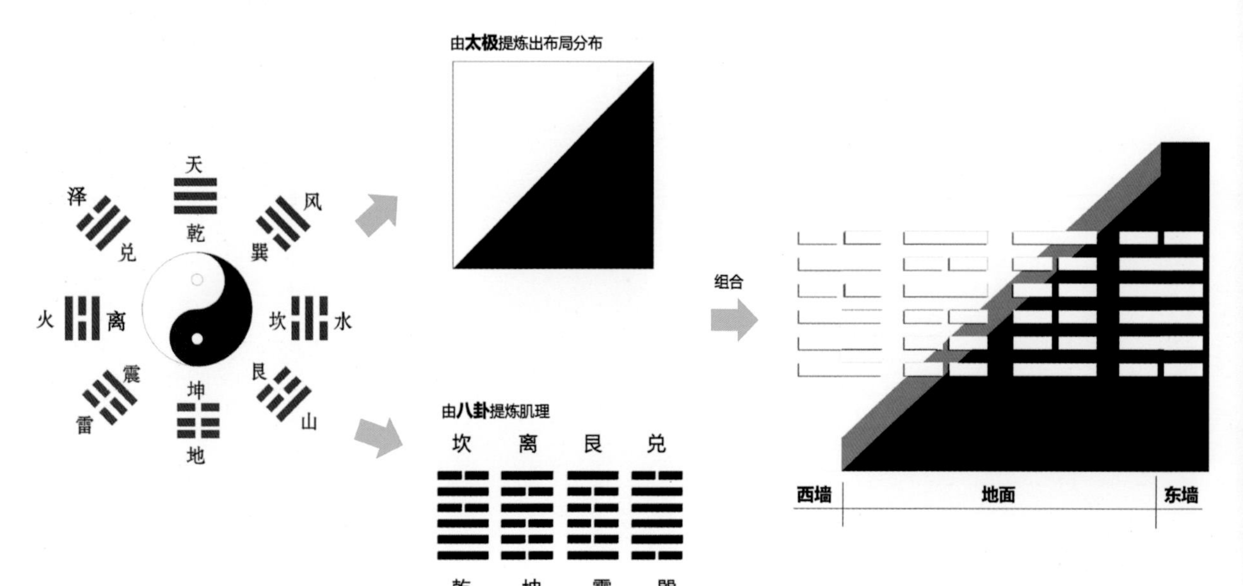

图 3-116 两仪之界设计思路
Figure 3-116 Design process of Garden of Tai Chi

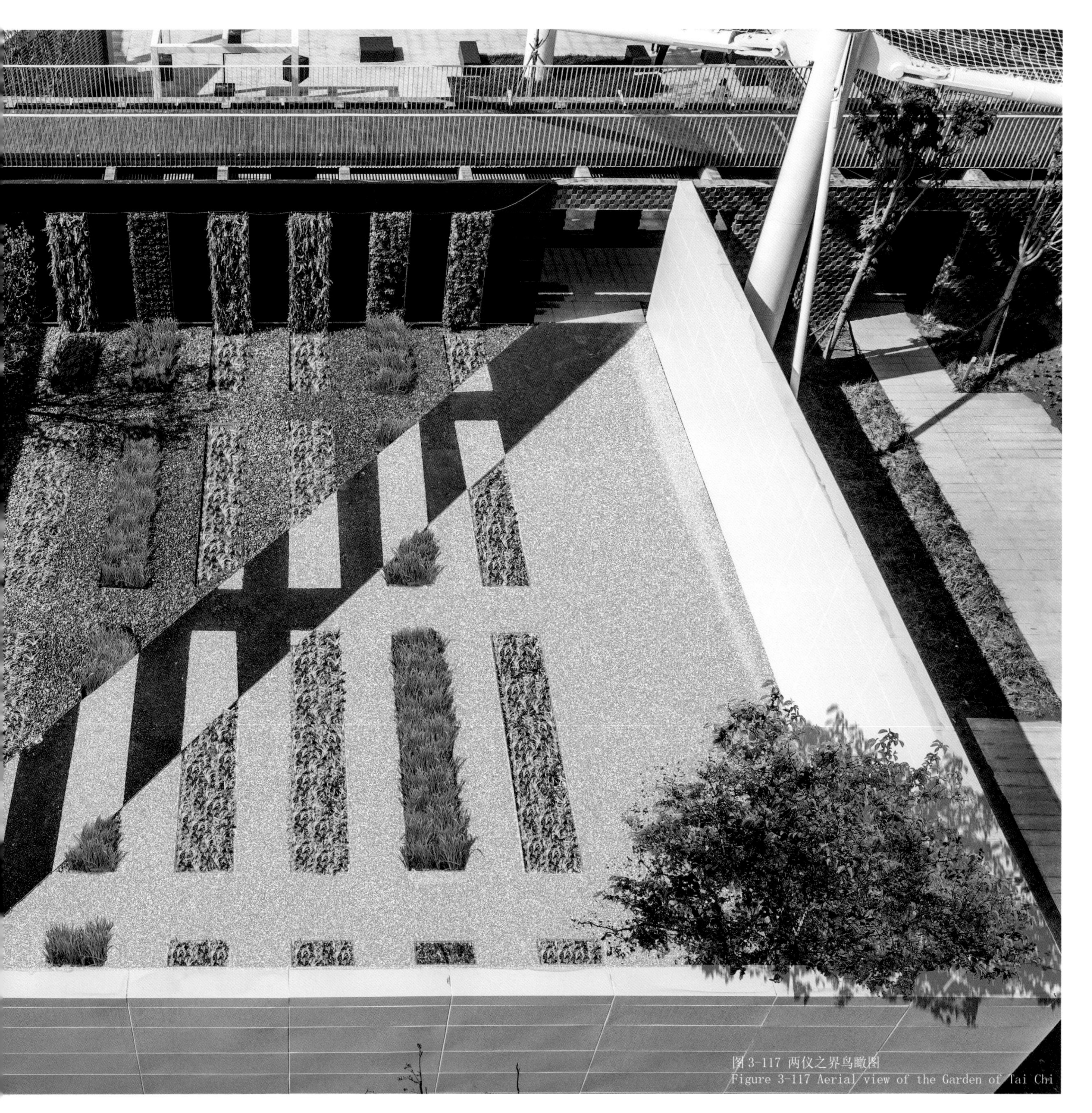

图 3-117 两仪之界鸟瞰图
Figure 3-117 Aerial view of the Garden of Tai Chi

两仪之界占地面积 900m^2，概念来源于阴、阳及八卦“乾、坤、震、巽、坎、离、艮、兑”。

The Garden of Tai Chi covers an area of 900 square meters. The design concept is derived from Yin and Yang, and the Eight Diagrams “Qian (sky), Kun (earth), Zhen (thunder), Xun (wind), Kan (water), Li (fire), Gen (mountain), and Dui (pond).”

一条黑色步道自东南向西北对角穿过，将空间分为一黑一白两个板块，用黑白两色碎石铺装指代阴阳两极，钢板墙面也被刷成了黑白两色，形成鲜明对比。以八卦的卦象为肌理形成的地面绿化及墙面立体绿化，给游客带来强烈的视觉冲击力，打造一个纯粹的道意空间。

A black trail passes the garden diagonally from southeast to northwest, dividing the space into two plates, namely the black and the white. The black and white gravel paving refers to Yin and Yang. The steel plate wall is also painted in black and white, demonstrating sharp contrast. The ground greening and the greening on the wall formed according to the Images of the Eight Diagrams give visitors a strong visual impact and create a pure Taoist space.

图 3-118 两仪之界
Figure 3-118 The Garden of Tai Chi

（10）出口展园

出口展园位于赵都新韵的尽端，面积约 450m²。场地东南和西北角共设计两处红色听筒廊架，廊架下部悬吊内置音箱、灯具等感应设备的红色玻璃钢罩，当游客将头伸进盒子内，设备会启动，与游客互动。

（10）Exit Garden

Exit Garden is located at the end of New Handan Pavilion, with an area of about 450 square meters. In the southeast and northwest corners of the site, there are two red earpiece gallery boxes. Red glass steel cover with built-in speakers, lamps and other sensing devices are suspended from the lower portion of the gallery. When visitors put their heads into the box, the equipment will start to interact with them.

此外，场地设有不规则方形种植池，种植池旁散落五种不同规格的红色玻璃钢坐凳，内嵌 LED 暖白光源。

In addition, the site is equipped with irregular square planting ponds, red glass fiber reinforced plastic benches of five different sizes are scattered around the planting ponds, in which LED warm white light embedded.

图 3-119 出口展园
Figure 3-119 Exit garden

3.2.6 景点三 山水邯郸

HANDAN PAVILION

山水邯郸主场馆总面积近 2 万 m^2，是园博会内规模最大、最为核心的标志性建筑，整个建筑由 4 个独立的场馆组成：1 号馆为植物馆，2 号馆为艺术馆，3 号馆为主场馆，4 号馆为会议中心，彼此之间通过立体廊桥和下沉庭院紧密相连，连绵山峦一般的巨型网架将 4 个场馆笼罩其下，共同构成一幅美轮美奂的建筑诗画，以山水之形反映园博园的设计主题。

The main hall of Handan Pavilion has a total area of nearly 20,000 square meters. It is the largest and most central landmark building in the park. The complex consists of 4 independent halls: Hall 1 is the Botanical Museum, Hall 2 the Art Museum, Hall 3 the Main Hall, Hall 4 the Conference Center. The four are closely connected by a gallery bridge and a sunken courtyard. The huge mountain-like framework structure covers the four spaces, forming a glamorous architectural image.

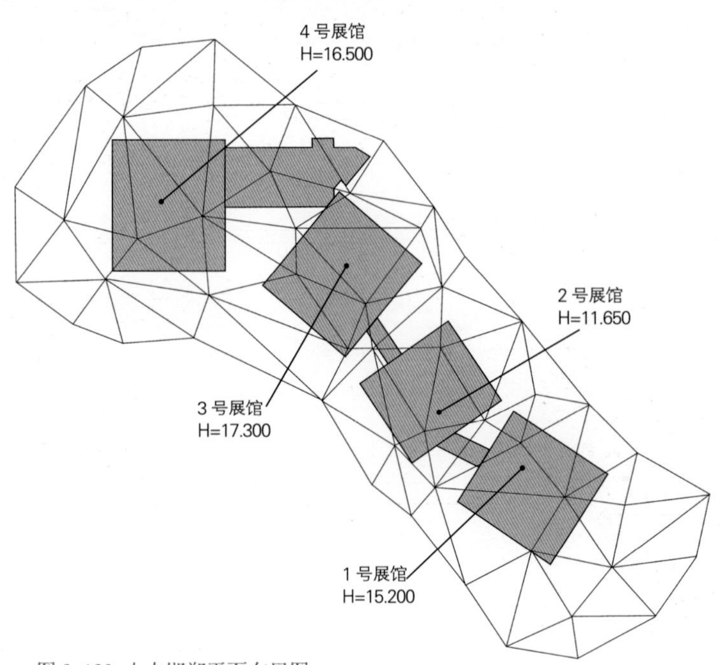

图 3-120 山水邯郸平面布局图
Figure 3-120 Architectural layout plan of Handan Pavilion

图 3-121 山水邯郸鸟瞰图
Figure 3-121 Bird eye view of Handan Pavilion

设计特色

DESIGN FEATURES

主场馆的设计构思紧扣“山水邯郸，绿色复兴”的主题，作为本届园博园标志性文化展览建筑，设计撷取峰峦叠嶂的巍巍群山作为母题，利用钢结构梁柱体系搭建出巨型空间网架，网架之上覆盖金属垂网，如缥缈的仙山坐落在绿色水畔，极具风情。山水邯郸建筑群与浮光揽月一起组成一片城市山林的浪漫图景：巨网之下，绿意盎然，人群熙攘，水畔生姿！

The design concept of the main hall is closely sticked to the theme of “Natural Handan, Green Revival”. As the iconic cultural exhibition building of the Garden Expo, the design captures peaks and ridges as the motif, and uses a steel structure beam—column system to build a huge space network, covered with a metal net, resembling a misty fairy mountain sitting on the bank of a clear water. Together with the Floating Moon Lake, Handan Pavilion buildings form a romantic picture of the urban landscape: the greenery, the crowds, and the waterfront, all sit under the giant net.

图 3-122 山水邯郸全景
Figure 3-122 Panorama of Handan Pavilion

4栋建筑单体功能清晰，形体方正，现代简洁，空中连廊与下沉庭院共同构成的立体交通空间，为游客提供了舒适宜人的半户外休憩场所。

Each of the four buildings has a clear function, and is designed in modern, simplified and square shape. The overhead gallery bridge and the sunken courtyard form a three-dimensional traffic space, providing visitors with a comfortable and pleasant semi-outdoor resting place.

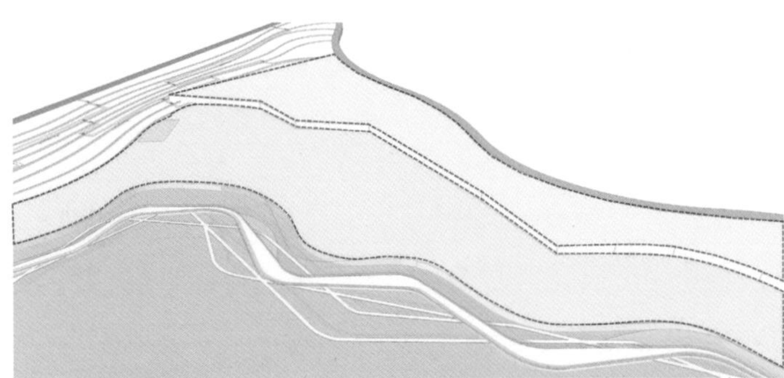

Step1 场地现状分析
场地和道路存在一定高差，西侧台田与场地相接，南部直面水雾森林，拥有良好景观面。

Step 1 Site current condition analysis
There is an elevational change between the site and its adjacent road. Where the west side is connected with terraced field, the south is featured with good forest scenery.

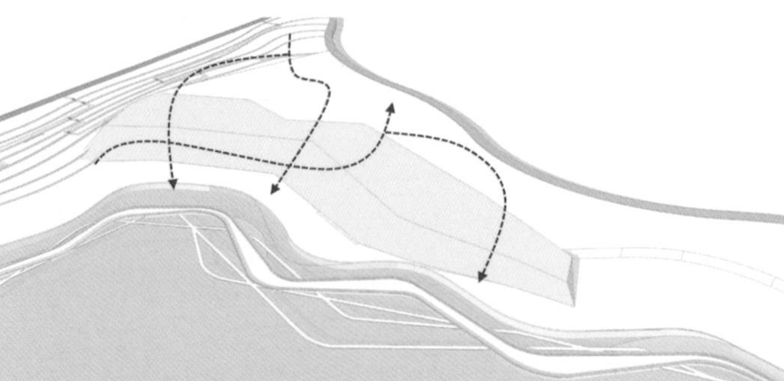

Step2 整理基底，与自然链接
利用自然缓坡将道路和水系连通，形成休闲娱乐的绿色场地。

Step 2 Reorganise existing green space, create human–nature relationship
Connect roads and water system through slopes to form a green recreational space.

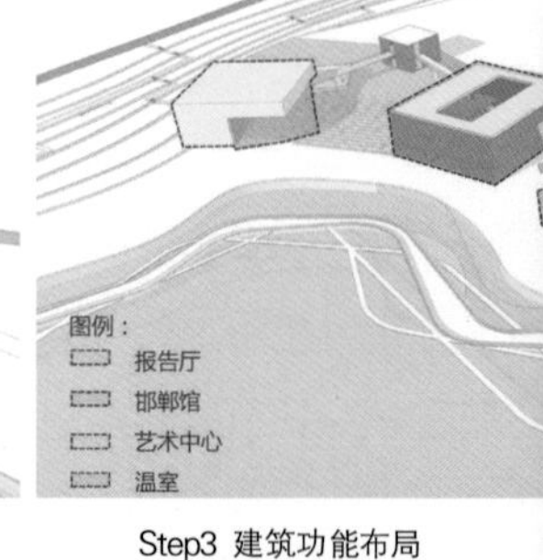

Step3 建筑功能布局
根据场地和道路方位布局山水邯
功能场馆，并采用连廊联系各场

Step 3 Architecture functional la
Develop functional layout in the
pavilion based on current site con
and road system. Use sky bridge
connect all the internal venues.

图 3-123 山水邯郸剖面图
Figure 3-123 Section of Handan Pavilion

图 3-124 概念分析
Figure 3-124 Conceptual analysis

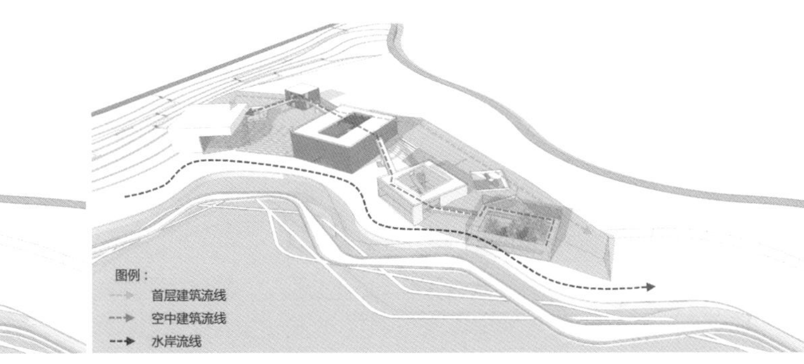

Step4 建筑流线分析
根据场地不同高差形成三条与自然相接的观展流线，丰富游客观展趣味性。

Step 4 Architecture circulation analysis
Take the advantage of elevational differences to form three visiting routes that are connected with green spaces thus enhance visiting experience.

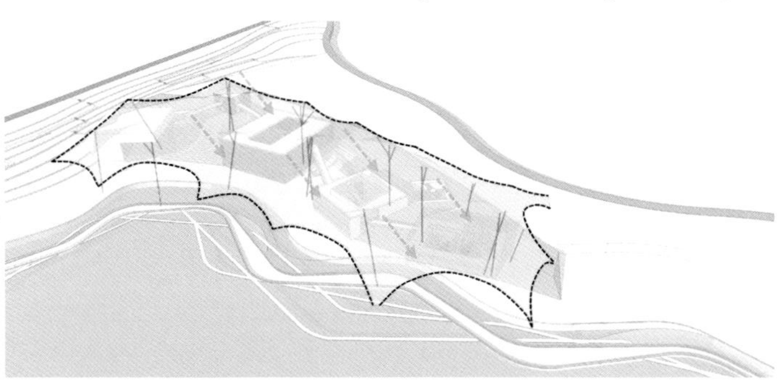

Step5 变化错落的遮阳设施
极具现代感的金属透光屋顶，形成场地特色标志性建筑，丰富场所感。

Step 5：Create the landmark of Handan Expo with modern metal shelter.
The modern metal shelter forms into a character landmark building and create the sense of place.

图 3-125 山水邯郸实体模型
Figure 3-125 Physical model of Handan Pavilion

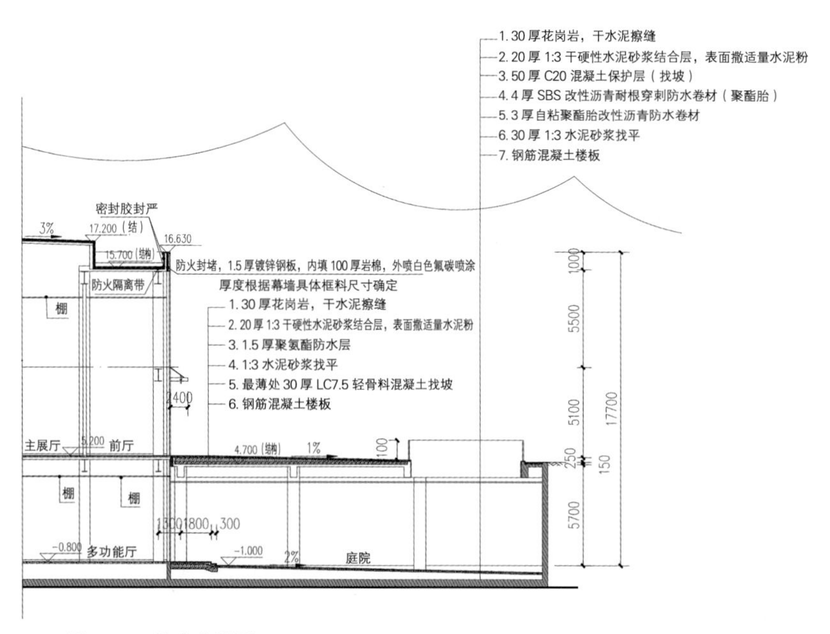

图 3-126 墙身大样图

Figure 3-126 Detail drawing of architectural wall

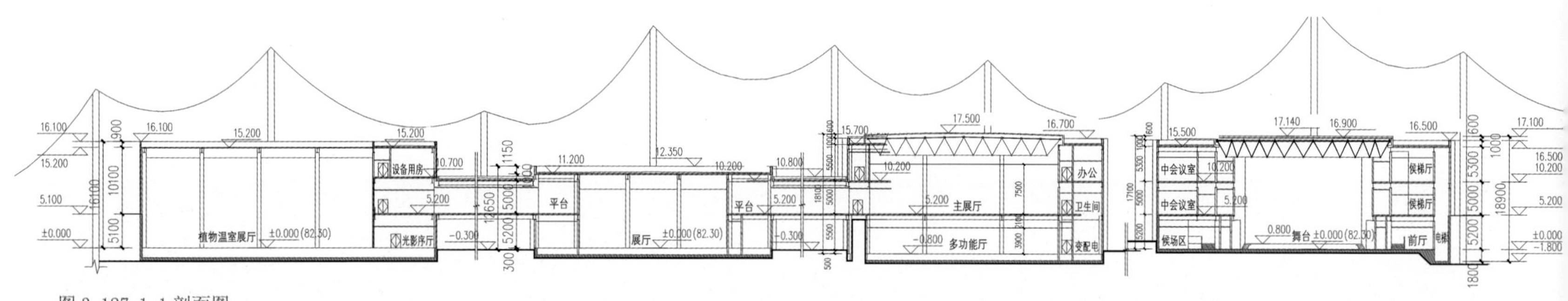

图 3-127 1-1 剖面图

Figure 3-127 Section 1-1

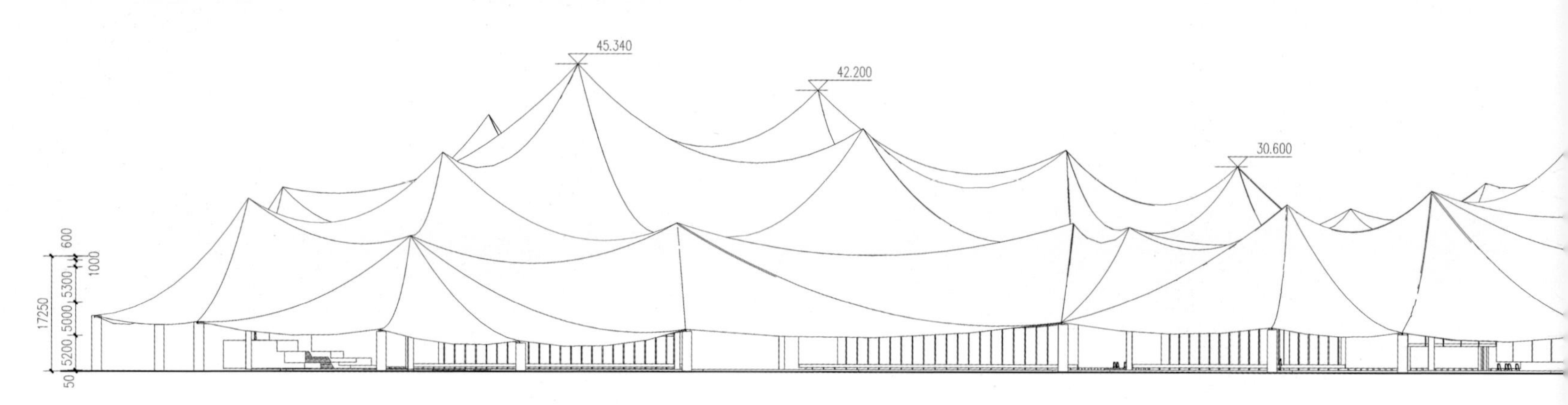

图 3-128 组合立面图

Figure 3-128 Composite elevation

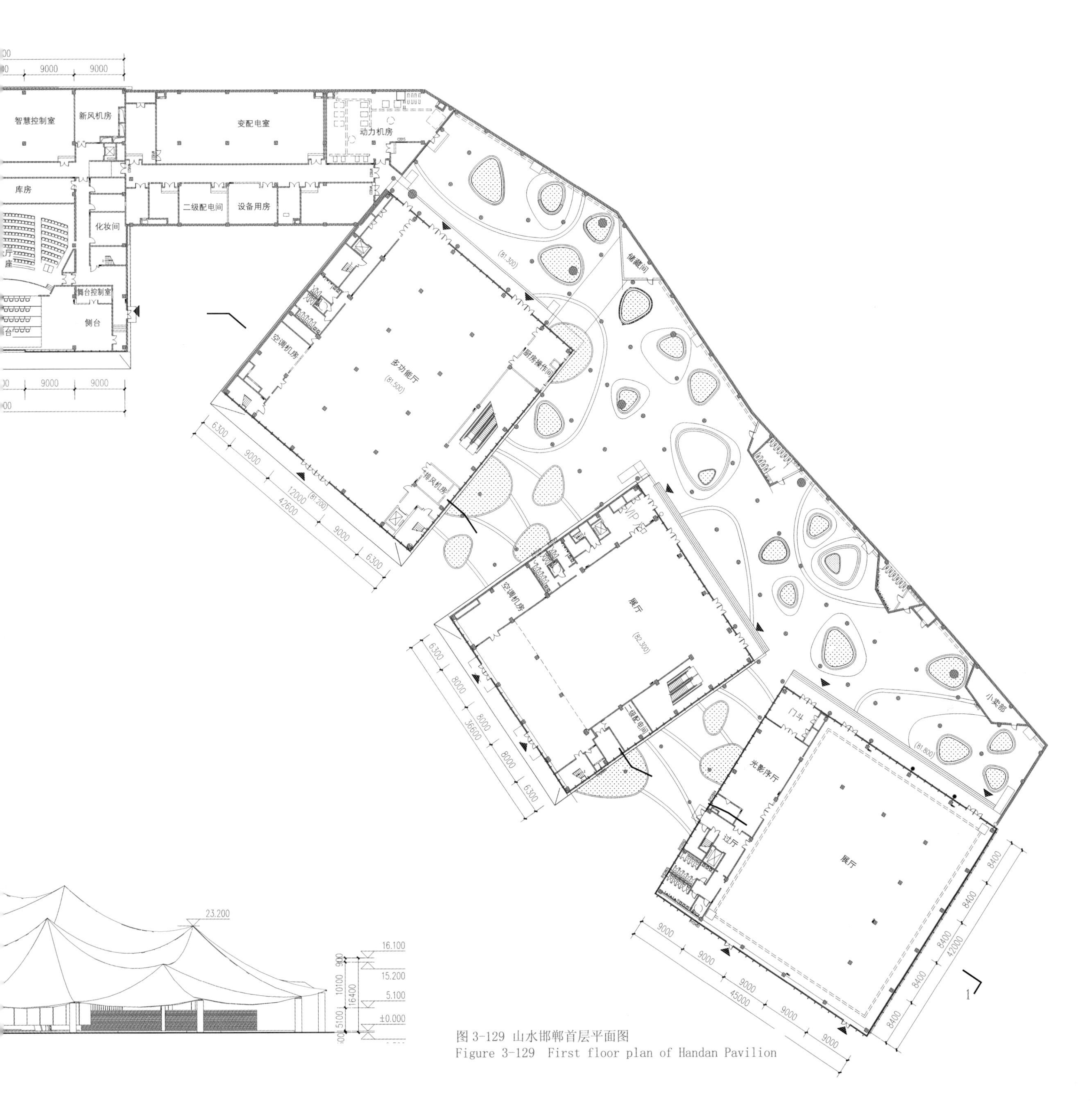

图 3-129 山水邯郸首层平面图
Figure 3-129 First floor plan of Handan Pavilion

图 3-130 山形网架
Figure 3-130 Mountain shaped framework

为了突出主场馆的标志性和文化性，设计在4栋功能建筑之上增设了一处连续的金属网，将整个建筑群落覆盖，100多根钢柱连同大跨度的曲线钢梁一起塑造出极具震撼力的连绵山体的形态。

In order to highlight the iconic and cultural nature of the main hall, a continuous metal net was added over the four functional buildings to cover the entire complex. More than 100 steel columns and large-span curved steel beams together form a stunning rolling mountain framework.

山形网架设计打破了传统建筑的结构体系，创新地采用固定几何形态与活动节点相组合的方式，在50m大跨度下仅用直径40cm的脊梁，最大限度地保证整体造型的轻盈状态；采用近5万m^2的连续金属网更是建筑史上的首例。同时在金属网上安装照明设备，当夜幕降临，华灯初上，在灯光照耀下的山水邯郸主场馆更加耀眼，成为邯郸的标志性景观和城市形象。

The mountain-shaped network design breaks the structural system of traditional architecture, and innovatively uses a combination of fixed geometric forms and movable nodes. For a span of 50 meters of network structure, spine beams as thin as 40 centimeters are used to maximize the lightness of the overall structure. The continuous metal net as large as 50,000 square meters is used for the first time ever in the history of architecture. Lighting equipment is installed on the metal net, when night falls, the main hall of Handan Pavilion under the light becomes the iconic landscape and the city image of Handan.

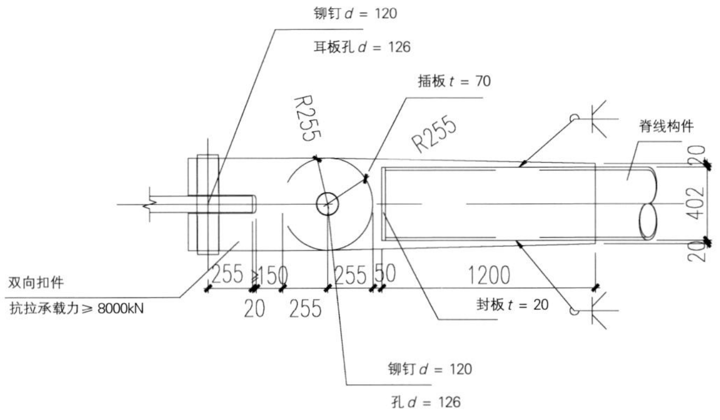

图3-131 1-1剖面做法图
Figure 3-131 Section 1-1

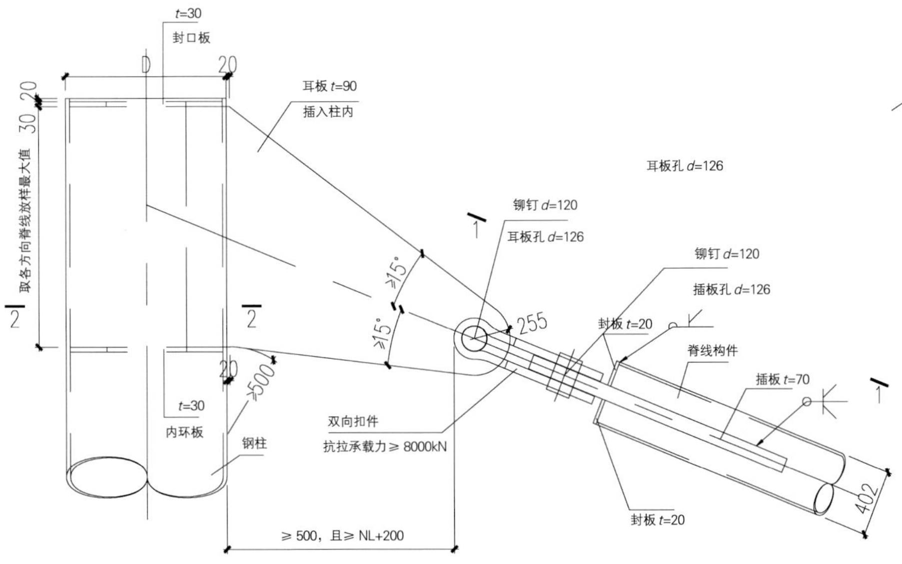

图3-132 钢柱与脊线构件连接节点大样图
Figure 3-132 Connection details between steel column and curved steel beam

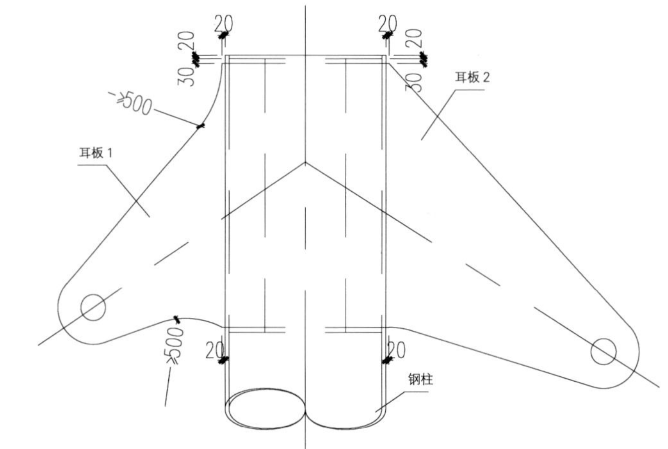

图3-133 不同角度和外伸长度的耳板做法大样图
Figure 3-133 Multi-angled extension joint details

图 3-134 沐浴晨光的山水邯郸
Figure 3-134 Handan Pavilion in the morning light

图 3-135 美轮美奂的建筑灯光秀
Figure 3-135 Glamorous architectural light show of Handan Pavilion

图 3-136 夜幕降临，华灯初上
Figure 3-136 Night lighting

图 3-137 园区“网红打卡地”之一
Figure 3-137 One of the Instagrammable checkin places on campus

3.2.7 景点四 青山画卷

FLOWERING TERRACES

齐村大坝是场地内重要的水利设施，高 5m 的陡峭混凝土护岸阻断了生态系统间的物质交换和能量循环，规则的几何形体消除了自然的异质性，水生植物与水生动物无处栖身，原有的自然生机消失殆尽。

青山画卷从中国古典山水画中吸取灵感，将大坝护岸改造为花田台地，形成宜游宜赏的连续景观空间，仿佛一幅展开的山水画卷，再次诠释园博园的主题。

The Qicun Dam is an important water conservancy facility in the site. The 5-meter-high steep concrete revetment blocks the material exchange and energy circulation between the ecosystems. The heterogeneity of nature is eliminated by the regular and geometric form. There is no habitat left for aquatic plants and aquatic animals, and therefore the original natural vitality disappears.

The Flowering Terraces draws inspiration from classical Chinese landscape paintings and transforms the dam revetment into a terraced field, forming a continuous landscape space suitable for both touring and appreciation. The whole garden resembles an unfolding landscape scroll, once again interpreting the theme of the park.

图 3-138 青山画卷现状实景
Figure 3-138 Photo showing existing condition of the Flowering Terraces

图 3-139 青山画卷鸟瞰图
Figure 3-139 Bird eye view of the Flowering Terraces

设计特色
DESIGN FEATURES

（1）台地消解堤坝陡峭高差

重塑齐村大坝地形，从中国古典山水画中吸取灵感，将原有陡坡层叠填土形成台地肌理，将大坝到揽月湖的高差转化为多层可利用的空间。台地共5级，台地间每层高差1m，宽窄不一，丰富的肌理变化为动植物的生存提供了可能。“青山画卷”与“浮光揽月”交相辉映，仿佛一幅展开的山水画卷。

（1）Dissolve Elevation Change of Dam with Terraces

Reshape the topography of the Qicun Dam, drawing inspiration from classical Chinese landscape paintings, forming a terraced texture by earth-moving on the original steep slopes, and transforming the elevation difference from the dam to the Floating Moon Lake into multiple layers of usable space. There are 5 levels of terraces, with height difference of 1 meter, and different widths. The rich texture change between the terraces provide habitats for a variety of species. Complemented by Floating Moon Lake, the Flowering Terraces look like an unfolded landscape scroll.

台地改造前
Before

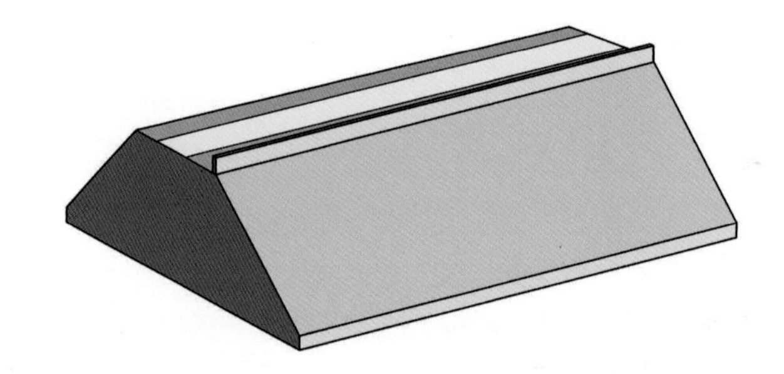

台地改造后
After

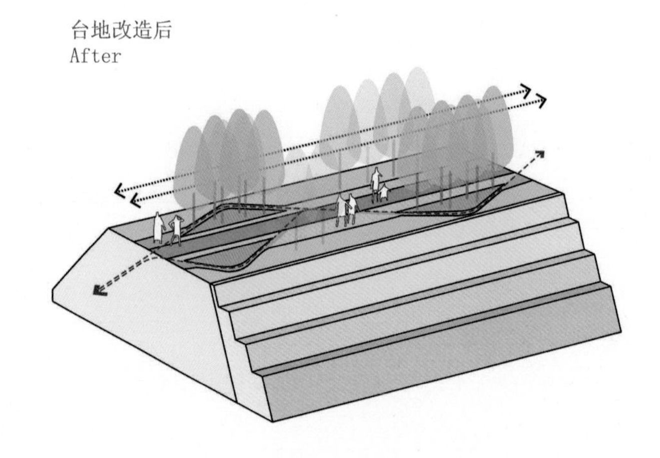

图 3-141 台地改造模式图
Figure 3-141 Illustrative diagram of land terracing

图 3-140 花涧瀑布
Figure 3-140 Terraced flower fields

（2）种植软化工程硬质界面

精心选择不同季节、不同花期的20余种花卉植物种植于台地中，四季之景不同，台间形态各异。种植设计以“杏临芳草、林隐青山”为主题，以特选山杏点缀观赏草、草花为基底形成特色景观，以黄山栾树等秋色叶树将小品半隐林中，实现景观的空间变化与季相变化。

（2）Planting Softens the Engineered Hard Surface

More than 20 kinds of flower plants are carefully selected to be planted on the terraces, representing the different scenery of the four seasons, and giving different patterns to the different levels of the terraces. The planting design is based on specially selected wild apricots embellished with ornamental grasses and flowers, and autumn leaf trees such as *Koelreuteria paniculata* to make the sketches half-hidden in the forest, enabling spatial changes and seasonal changes of the landscape.

花台乔木种植以果树为主，景观休息亭附近配置树形优美的特选大山杏，地被以蓝花鼠尾草、特色月季为主，配置火星花、蛇鞭菊、细叶美女樱等，形成条状花田，仿佛花涧瀑布，行云流水在花的海洋。

Fruit trees are the major tree selection of the terrace. Special wild apricots in beautiful shape are planted near the lookout rest boxes. The groundcovers used include *Salvia farinacea* and special Chinese rose, complemented by *Crocosmia crocosmiflora*, *Liatris spicata*, *Glandularia tenera*, etc. to formed a strip-shape flower field, looking like a waterfall flowing in the sea of flowers.

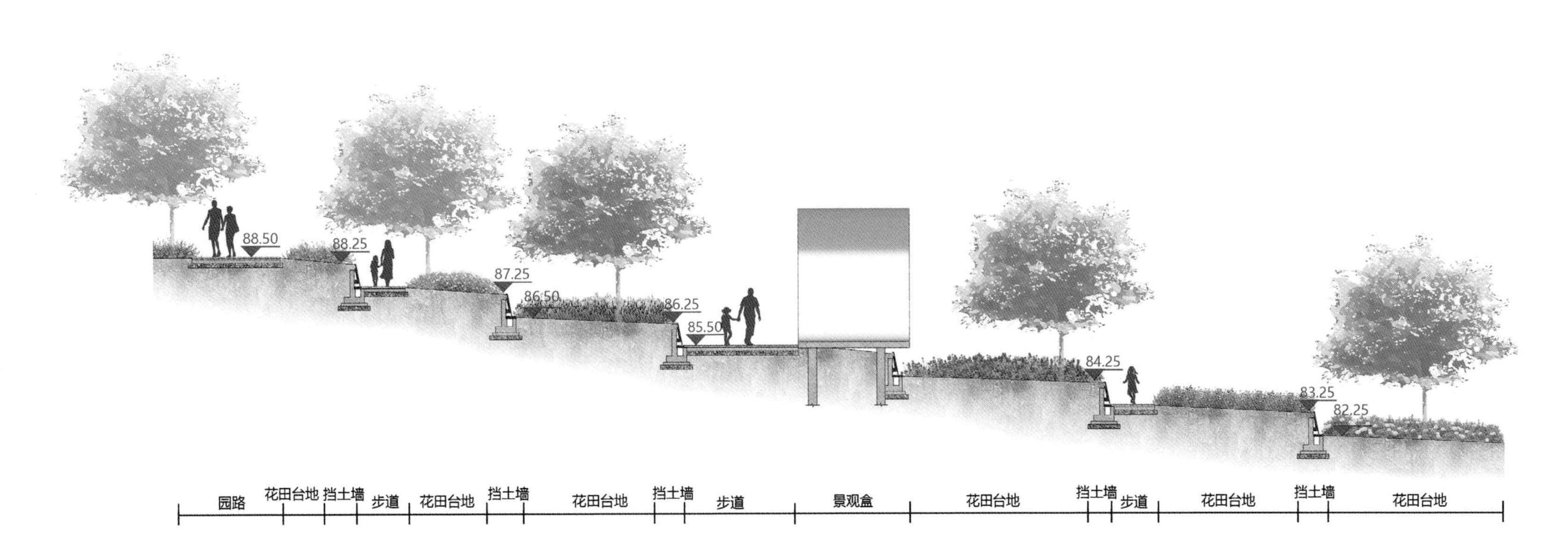

图3-142 青山画卷剖面图
Figure 3-142 Section of Terraced Flower fields

图 3-143 景观休憩盒
Figure 3-143 Lookout pergola

图 3-144 花田中点缀景观休息亭
Figure 3-144 Lookout pergola sitting in the terraced plant fields

（3）构筑物提供观景多样角度

台地花田间布置多处景观休息亭。景观休息亭形似卷云，采用白色玻璃钢材质，立面边界构成不同角度的框景画面。

亭子角度不同，亭内风景各异，以最少的笔墨实现了功能与设计美的统一。

（3）Structures Provide Various Lookout Angles

Multiple lookout boxes are arranged among the terraced flower fields. The lookout boxes are shaped like wrapped clouds, and are made of white glass fiber reinforced plastic. When looking out, the facade boundary of the boxes forms framed scenes of different angles.

According to the different angles of the lookout boxes, the scene seen from each box looks different, too. The functional value and design beauty here are unified by minimum effort.

图 3-145 耐候钢板种植池
Figure 3-145 Weathering steel planter

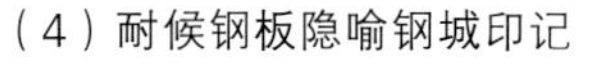

（4）耐候钢板隐喻钢城印记

花台栈道间利用多处不规则多边形耐候钢板种植池、花箱及挡墙遮挡混凝土坝体，既不破坏大坝的结构，又可作为栏杆保障通行安全。

（4）Weathering Steel Metaphorizing the Steel City Imprint

A number of irregular weathering steel planters, flower boxes and retaining walls are used among the boardwalks of the flower terraces to block the body of the concrete dam, which will not damage the structure of the dam, while can be used as a railing to ensure traffic safety.

3.2.8 景点五 工业遗风

INDUSTRIAL HERITAGE PAVILION

工业遗风所在区域原是邯钢的水渣场，场地遗留了巨大的矿粉罐和部分工业厂房，设计将其改造为园区内的用餐区，建筑面积 2070.18m^2。

The area where Industrial Heritage Pavilion is located was formerly the water slag field of Handan Steel. A huge mineral powder tank and part of the industrial plant were left on site, which was transformed into a canteen with a construction area of 2070.18 square meters.

图 3-146 工业遗风现状实景
Figure 3-146 Photo showing existing condition of the Industrial Heritage Pavilion

图 3-147 工业遗风鸟瞰图
Figure 3-147 Bird eye view of the Industrial Heritage Pavilion

设计特色

Design Features

设计在原有工业塔及保留结构的基础上对其进行改造，丰富建筑空间、增加建筑面积以满足园博会对餐饮建筑的需求，并在不同体量的建筑之间增加连廊及阴篷，为游客在疲惫的时候送来一片阴凉惬意的悠闲空间。

By transforming the original industrial tower and the retained structure, the design enriches and increases the building area to meet the needs of the Garden Expo for catering buildings. It also increases corridors and shade sheds between the buildings, thanks to which, visitors are sent a cool and comfortable recreation space.

设计用透明 LED 屏幕将原有储料罐进行外立面电子化处理，夜晚可展现无限可能的创意图案；同时采用变化的白色格栅对保留结构进行立面肌理的重新设计，加建必要的交通联系空间及生态室外廊架，创造宜人的半户外空间。

The design uses transparent LED screens to make the original storage tanks electronical, so that creative patterns can be displayed on them at night. A changing white grille is used to redesign the facade texture of the retained structure. Traffic connection space and ecological outdoor corridors are added to create a pleasant semi-outdoor space.

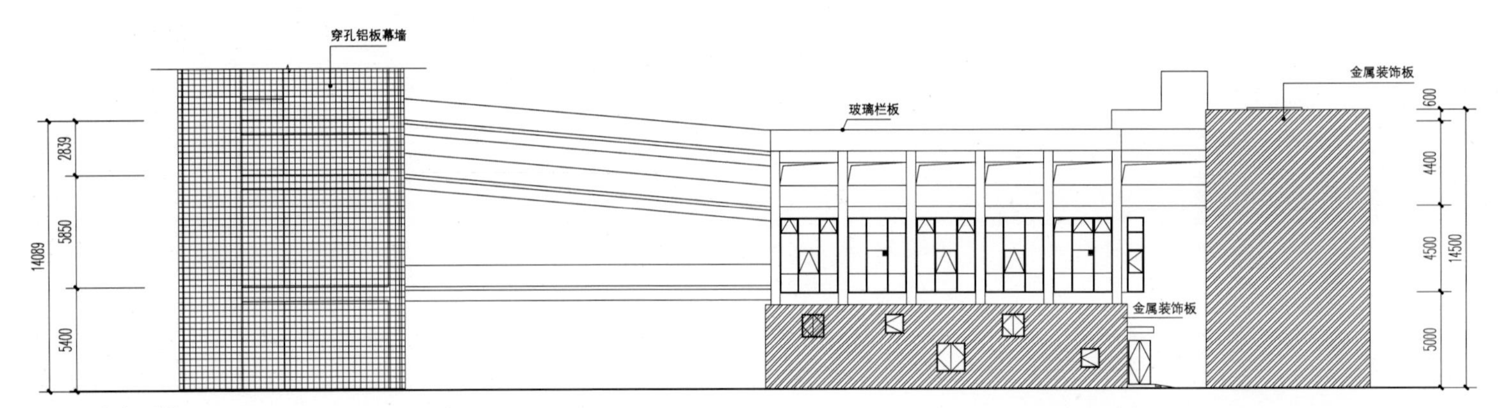

图 3-148 东立面图
Figure 3-148 East elevation

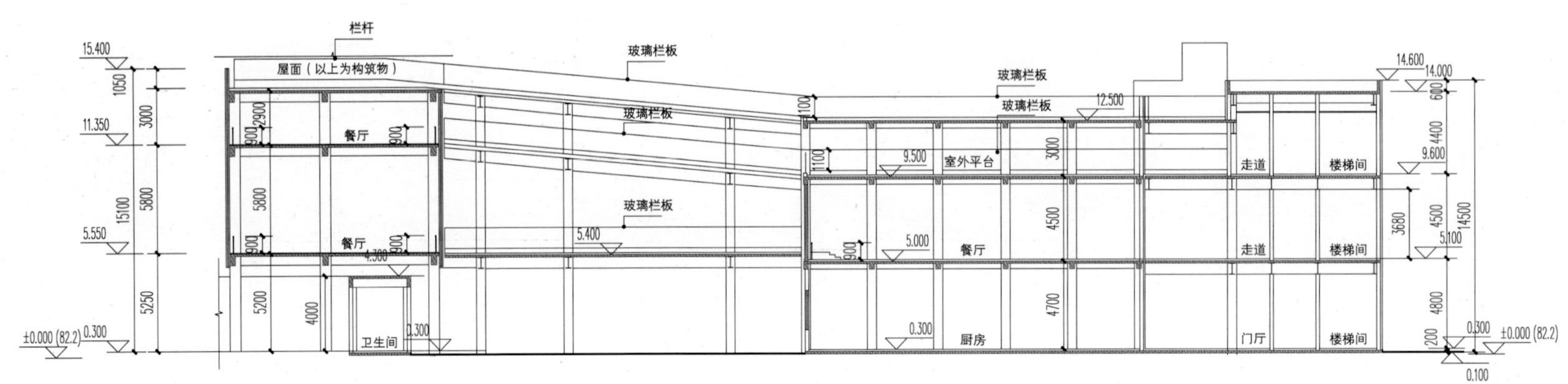

图 3-149 剖面图
Figure 3-149 Section

图 3-150 工业遗风效果图
Figure 3-150 Industrial Heritage Pavilion rendering

图 3-151 建筑改造剖透图
Figure 3-151 Sectional perspective drawing of the architectural renovation

图 3-152 工业遗风施工过程
Figure 3-152 Construction process of the Industrial Heritage Pavilion

图 3-153 户外空间
Figure 3-153 Outdoor space

3.2.9 景点六 清渠如许

WETLAND TERRACES

清渠如许改造前是场地中微高起的土丘，植被稀少，仅有零散的几株灌木；土壤质量较差，薄薄一层表层土壤下方填满了各类工业矿渣和建筑垃圾。

设计团队运用中国传统农业中的“梯田”和“陂塘”形式，结合科学的生态净化方法，将城市排放的废弃中水引向山顶，再经过台地植物层层净化，最终成为满足景观使用需求的水体。

清渠如许是整个园区的生态净化展示区，充分体现了“绿色复兴”的园博园设计理念。

Before the reconstruction, the Wetland Terraces site was a slightly raised mound area, with only a few scattered shrubs. The soil quality of this area was poor and, under the thin layer of the surface soil there was various industrial slag and construction waste.

The design team uses the “terraced fields” and “ponds”, forms of traditional Chinese agriculture, combined with scientific ecological purification methods, to lead the waste water discharged from the city to the top of the mountain. Then the water is purified by the terrace plants layer by layer for landscape use.

The Wetland Terraces is the ecological purification exhibition area of the entire park, which fully embodies the “green revival” design concept of the park.

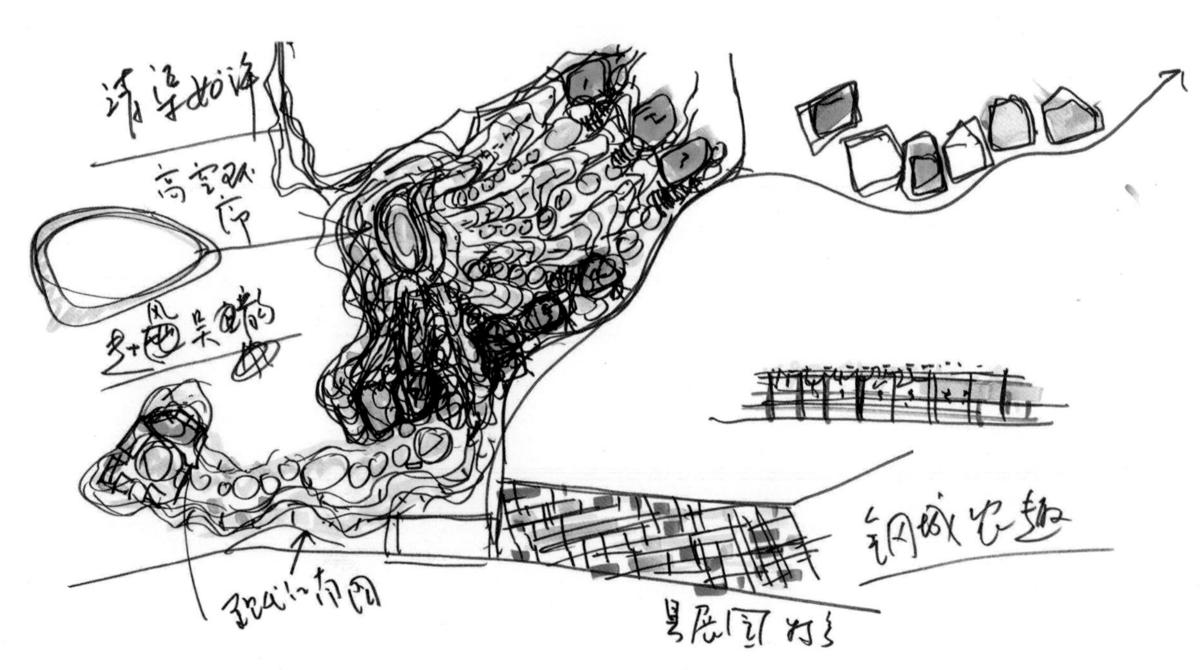

图 3-154 首席设计师俞孔坚清渠如许节点草图
Figure 3-154 Sketch drawing of Wetland Terraces by chief designer Yu Kongjian

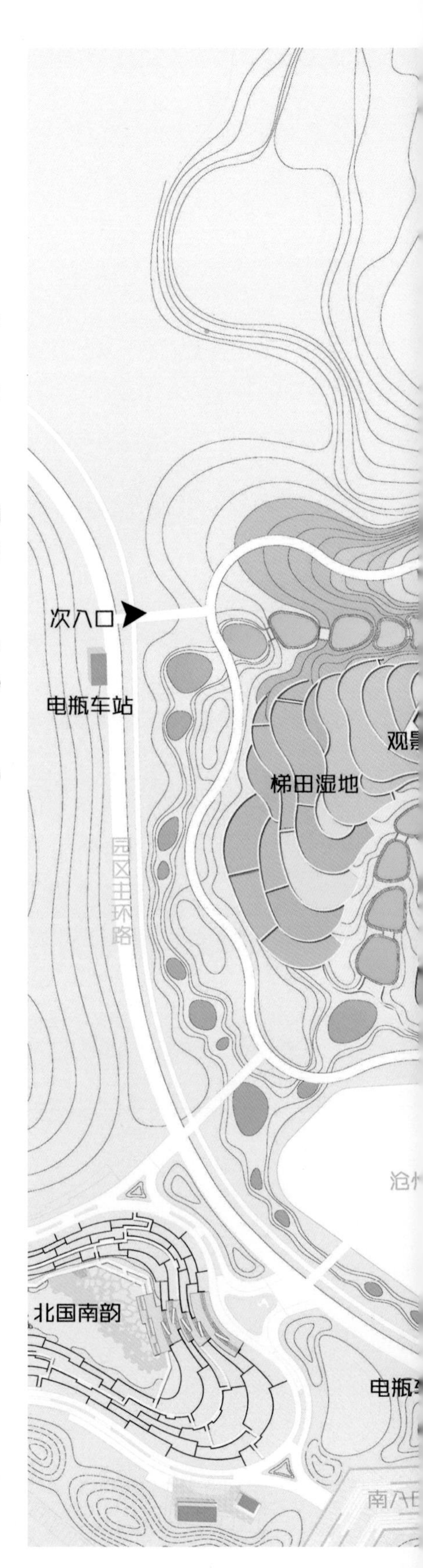

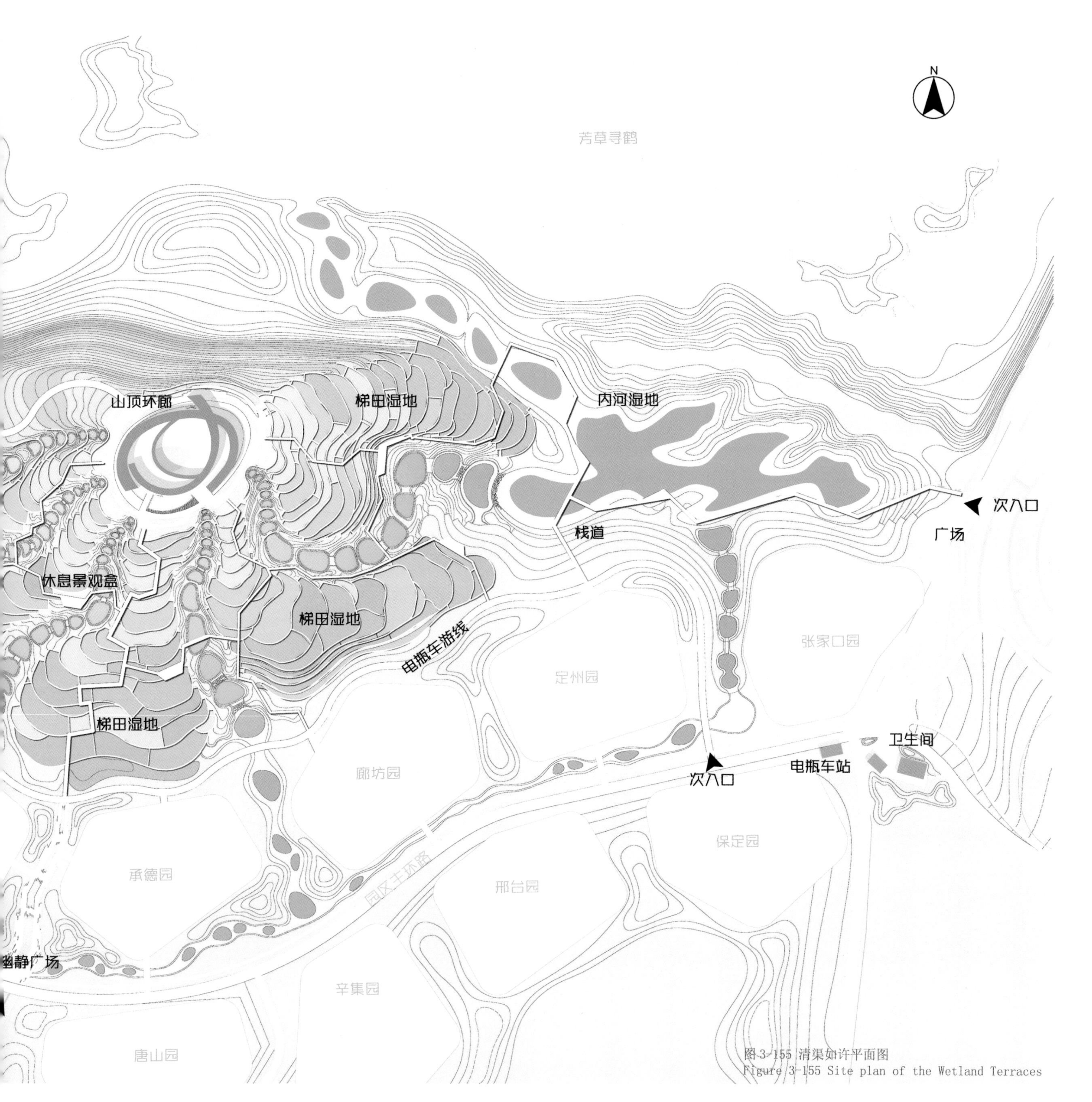

图3-155 清渠如许平面图
Figure 3-155 Site plan of the Wetland Terraces

设计特色

DESIGN FEATURES

（1）生态湿地的形式——“梯田”和“陂塘”

湿地的形式设计为田和塘两种，自山顶至山脚，分别形成连续的5条梯田带和5条陂塘带。市政中水自山顶流入各条梯田带和陂塘带中，由湿地中的植物、微生物、基质等净化水体中的营养物质，层层跌落的水流不但给水体补充了氧气，也营造了活泼的景观氛围。最终，经过净化的水汇入山脚的内河，供人们嬉戏、观赏。

（1）Ecological Wetland Forms—“Terraced Fields” and “Ponds”

Wetlands are designed in two forms: fields and ponds. From the top of the mountain to the foot of the mountain, five continuous terrace belts and five pond belts are formed respectively. Municipal reclaimed water flows from the top of the mountain into the terraces and ponds. The plants, microorganisms, and substrates in the wetland are used to purify the nutrients in the water body. The falling water stream not only supplements the water body with oxygen, but also creates a lively landscape atmosphere. Finally, the purified water flows into the inland river at the foot of the mountain for visitors to play and appreciate.

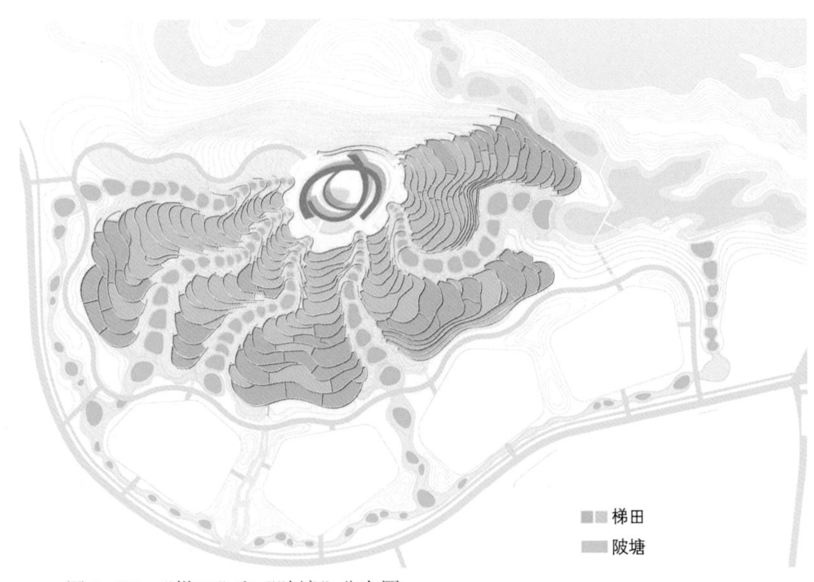

图3-156 “梯田”和“陂塘”分布图
Figure 3-156 Layout plan of “terraced fields” and the “ponds”

图3-157 清渠如许鸟瞰图

Figure 3-157 Aerial view of Wetland Terraces

（2）生态湿地的净化流程——4 种湿地组合循环

综合考虑生态湿地的形式、场地高差、水质等要素，设计团队选取了 4 种湿地，分别是表流湿地、水平潜流湿地和上行垂直流湿地，进行净化。

氧化塘深度一般不超过 1.5m，阳光可以直接射透到塘底，主要由塘内的微生物、水生植物对有机物进行降解。

表流湿地深度一般不超过 0.5m，种植大量挺水植物，它们为微生物提供了大量的附着空间；微生物的好氧反应、植物过滤和吸附等过程共同实现水体净化。

水平潜流湿地：水流在地表下流动，充分利用了填料表面生长的生物膜和丰富的植物根系；同时保湿性又好，处理效果受气候影响小；水力负荷和污染负荷大，对 BOD、COD、SS 和重金属等污染指标的去除效果好。

上行垂直流湿地：水流由底部流向上部；主要通过植物吸收、基质吸附和硝化—反硝化作用去除污染物，去除氨氮和磷的效果好。上行垂直流湿地的硝化能力高于水平潜流人工湿地，可用于处理氨氮含量较高的污水，但是其有机物的去除能力不如水平潜流湿地，因此二者经常组合形成复合湿地。

（2）The Purification Procedure of Ecological Wetland—Four Types of Wetland Combination Cycle

Considering factors such as the form of ecological wetland, site height difference, water quality etc., the design team adopted four types of wetlands for purification, namely surface wetland, subsurface wetland, horizontal subsurface wetland and vertical subsurface wetland.

Depth of oxidation ponds is generally no more than 1.5 meters, so that sunlight can directly reach to the bottom of the pond. It mainly depends on the microorganisms to degrade organics and aquatic plants.

Depth of surface wetland is generally no more than 0.5 meters. A large number of emergent plants are used, which provide sufficient adsorption space for microorganisms. The aerobic reaction of microorganisms, plant filtration and adsorption processes together help achieve water purification.

The water flows under the surface, making full use of the surface of filteration and enrich plant roots; the moisture retention is good, and the treatment effect is less affected by the climate; the hydraulic load and pollution load are huge. and with good removal effect of BOD, COD, SS and heavy metals.

Water flows upwards from the bottom to top. The subsurface wetland mainly removes pollutants through plant absorption, substrate adsorption and nitrification–denitrification. It has better removal effect of ammonia nitrogen and phosphorus the other two types of wetland, the vertical subsurface wetland works better on removing organic matters. Hence the two always integrate into a composite wetland.

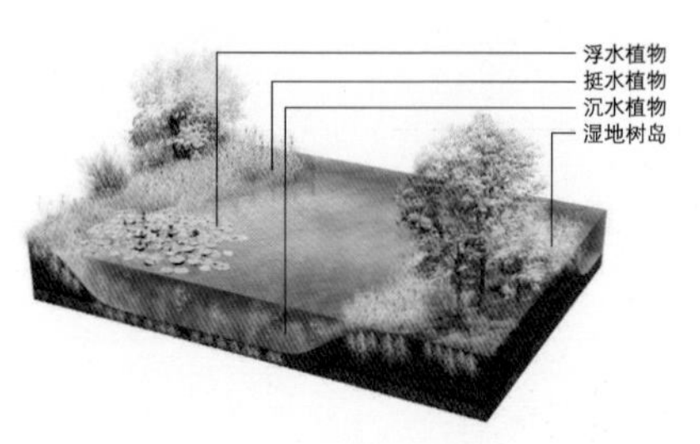

图 3-158 氧化塘模式图
Figure 3-158 Model diagram of oxidation pond

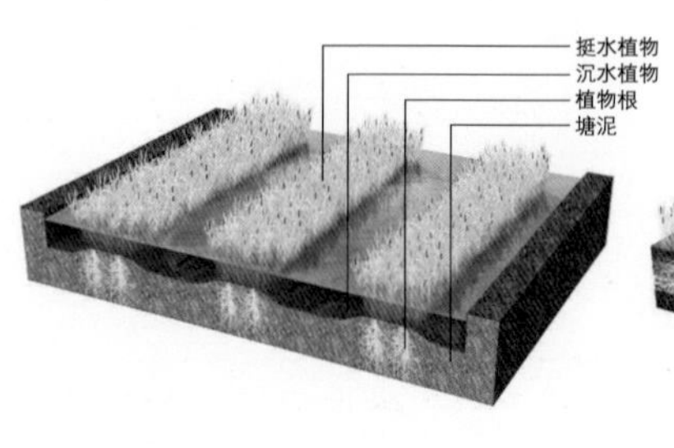

图 3-159 表流湿地模式图
Figure 3-159 Model diagram of surface wetland

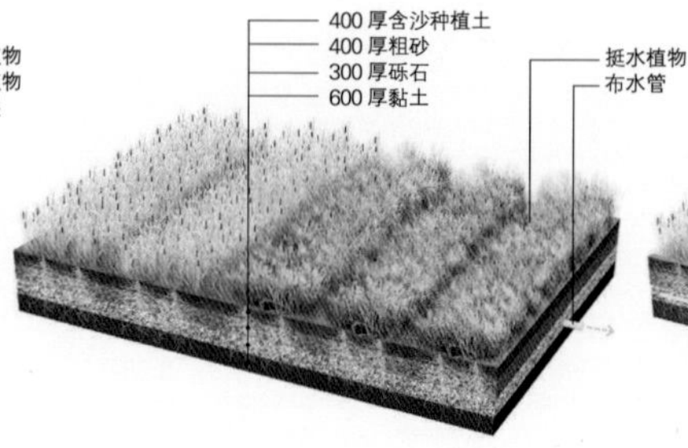

图 3-160 水平潜流湿地模式图
Figure 3-160 Model diagram of horizontal subsurface wetland

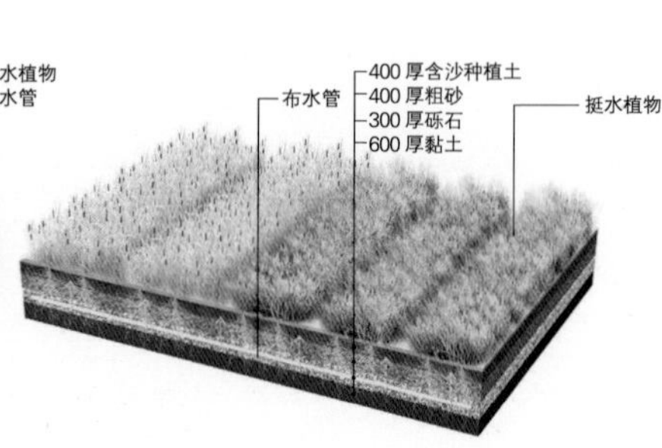

图 3-161 上行垂直湿地模式图
Figure 3-161 Model diagram of vertical subsurface wetland

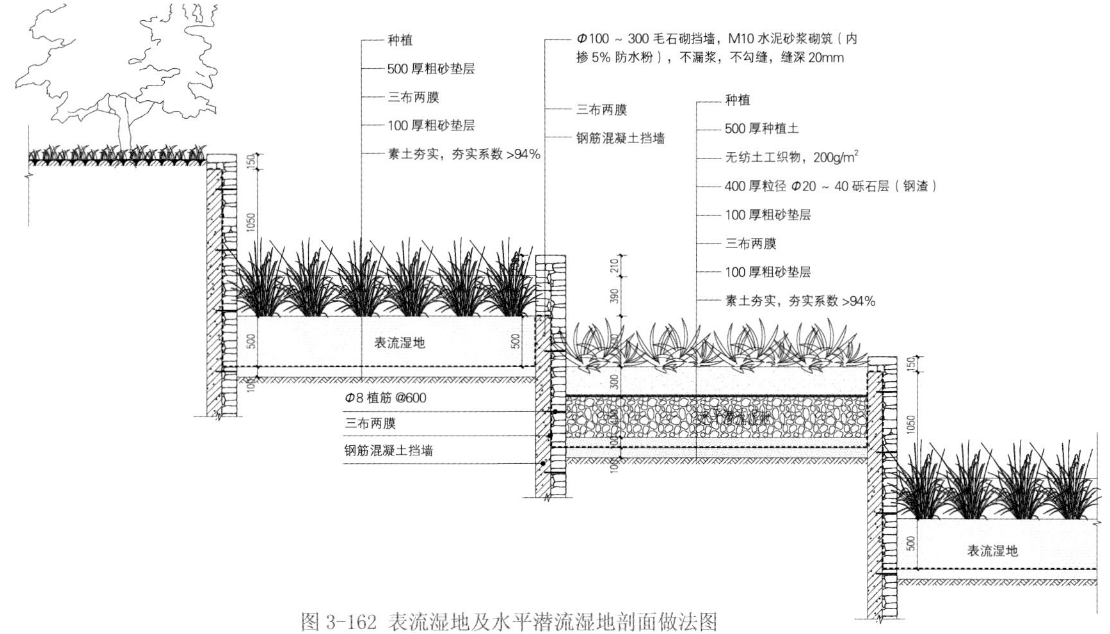

图 3-162 表流湿地及水平潜流湿地剖面做法图

Figure 3-162 Section of surface wetland and subsurface horizontal-flow wetland

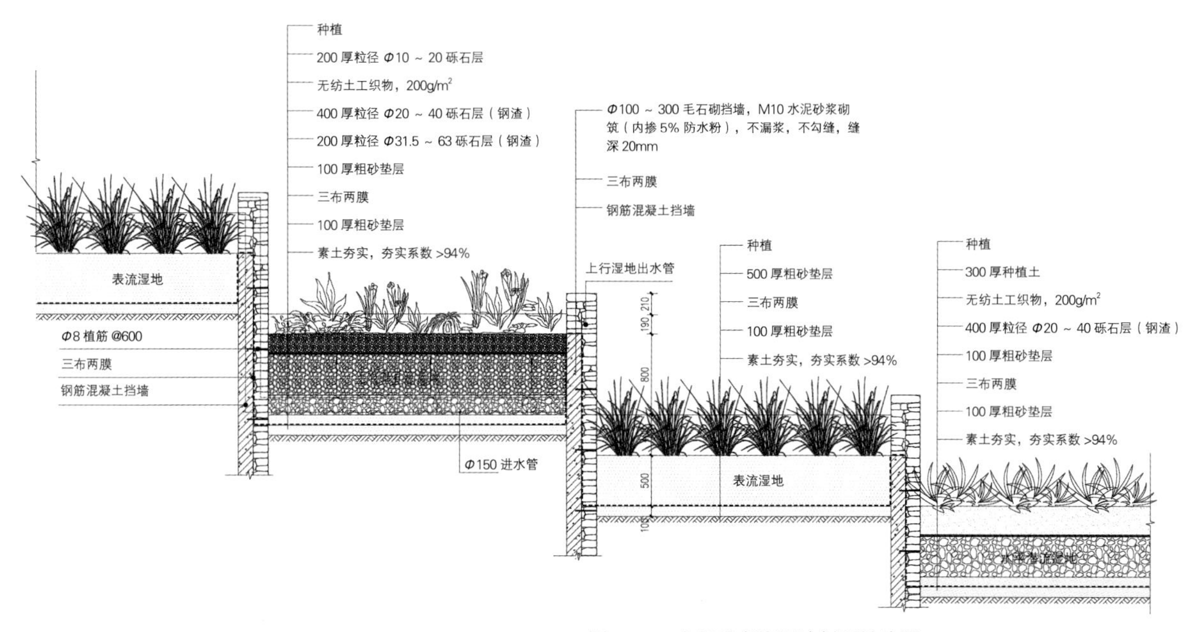

图 3-163 上行垂直流湿地剖面做法图

Figure 3-163 Section of upward vertical surface wetland

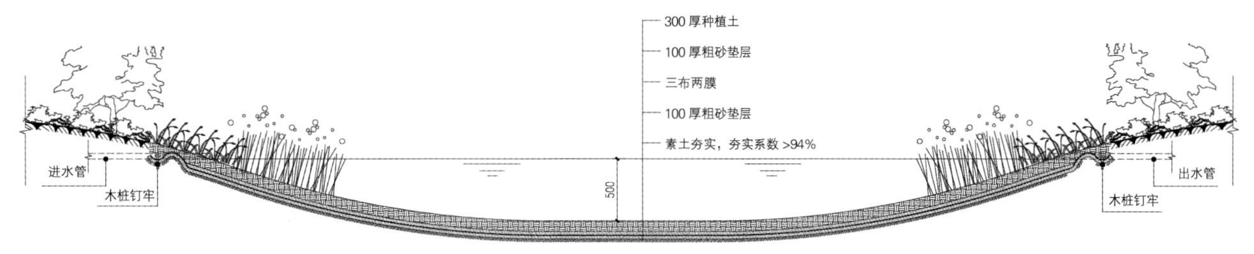

图 3-164 氧化塘做法图

Figure 3-164 Section of oxidation pond

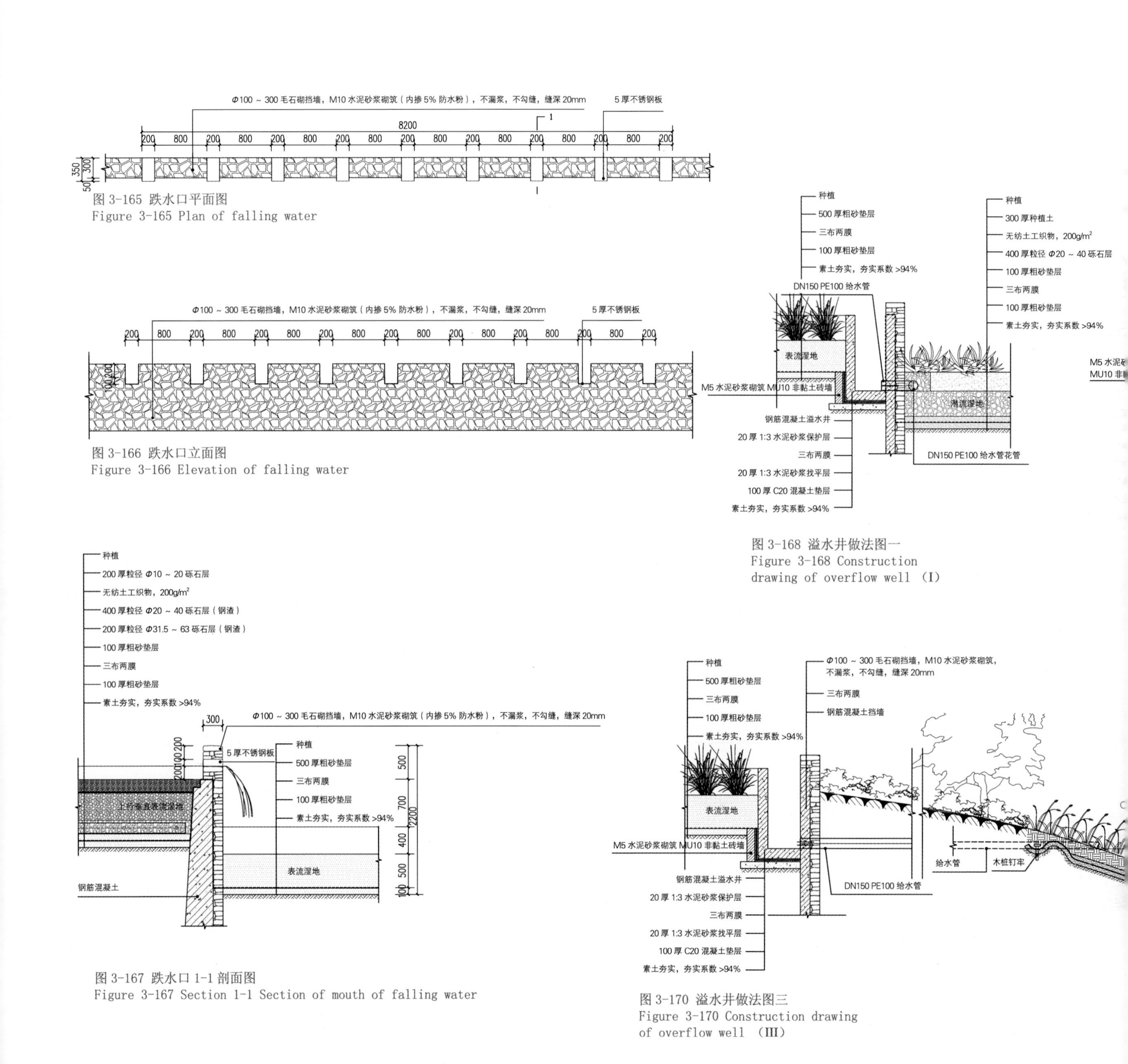

图 3-165 跌水口平面图
Figure 3-165 Plan of falling water

图 3-166 跌水口立面图
Figure 3-166 Elevation of falling water

图 3-168 溢水井做法图一
Figure 3-168 Construction drawing of overflow well （I）

图 3-167 跌水口 1-1 剖面图
Figure 3-167 Section 1-1 Section of mouth of falling water

图 3-170 溢水井做法图三
Figure 3-170 Construction drawing of overflow well （III）

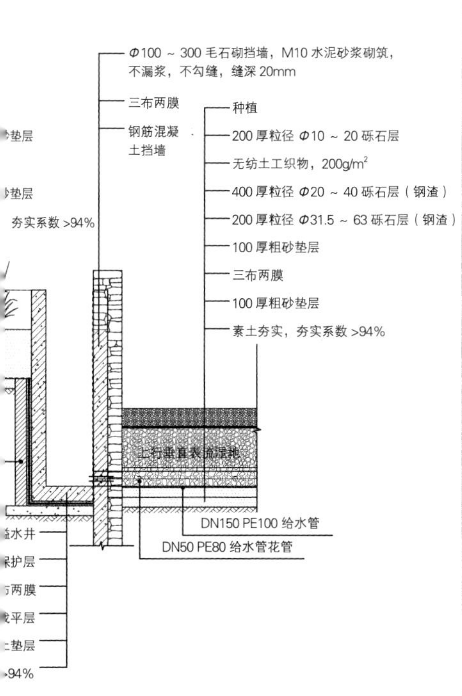

图 3-169 溢水井做法图二
Figure 3-169 Construction drawing of overflow well （II）

图 3-171 防渗施工过程
Figure 3-171 Waterproof construction process

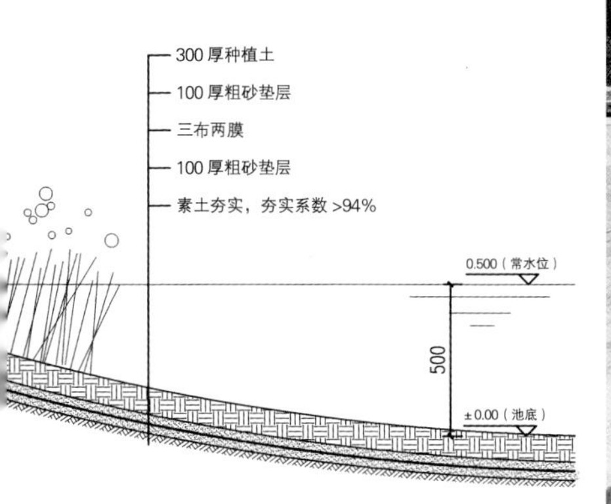

图 3-172 “三布两膜”施工过程
Figure 3-172 Constriction process of waterproof layers

4 种湿地在净化原理和净化能力上各有差别，对这 4 种湿地进行一定的排序和组合，就可以更好地发挥净化作用，这种排序和组合就是净化流程。清渠如许采用氧化塘—表流湿地—氧化塘—潜流湿地的基本净化流程单元，在整个场地中多次重复采用这一流程，以强化净化效果。

The four types of wetlands differ in their purification principles and capabilities. By certain sorting and combination of the four types, we can achieve better purification effect. This sorting and combination is called the purification procedure. The Wetland Terraces adopts a basic purification procedure unit of oxidation pond—surface wetland—oxidation pond—subsurface wetland, and reproduces this procedure repeatedly throughout the site.

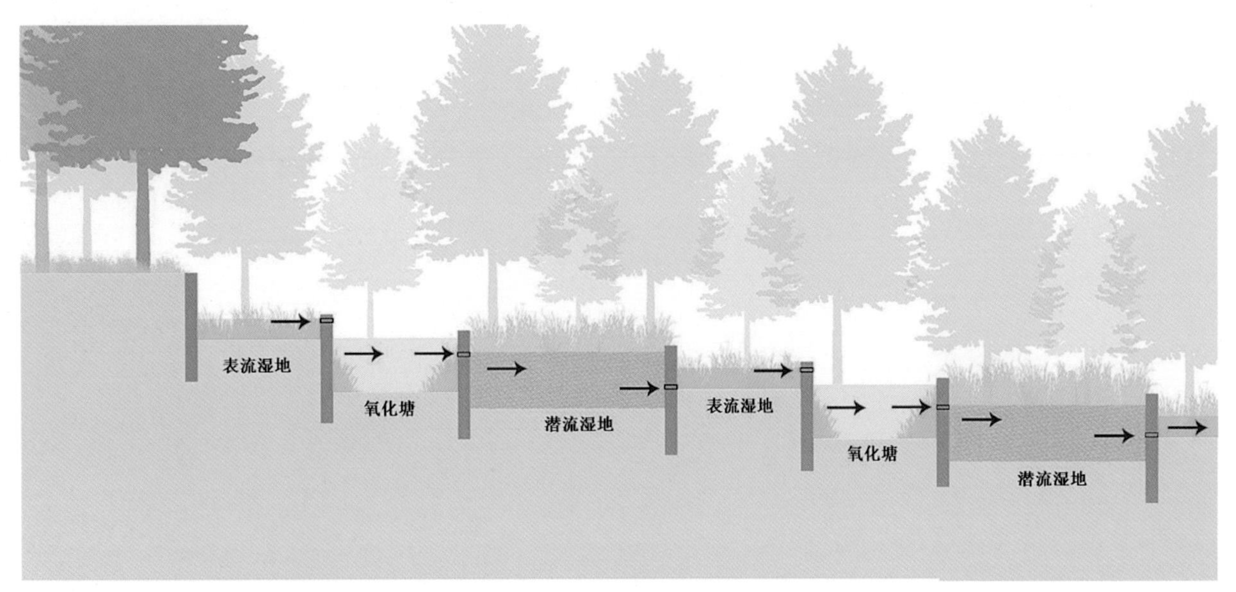

图 3-173 净化流程剖面示意图
Figure 3-173 Illustrative section of the water purification procedure

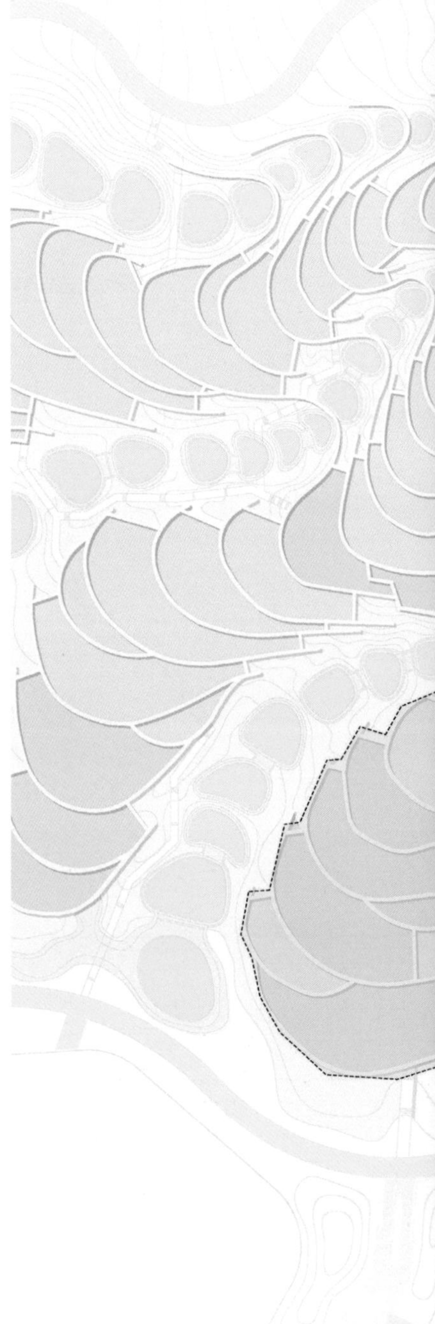

图 3-174 净化工艺模块选取示意图
Figure 3-174 Illustrative plan of the water purification procedure

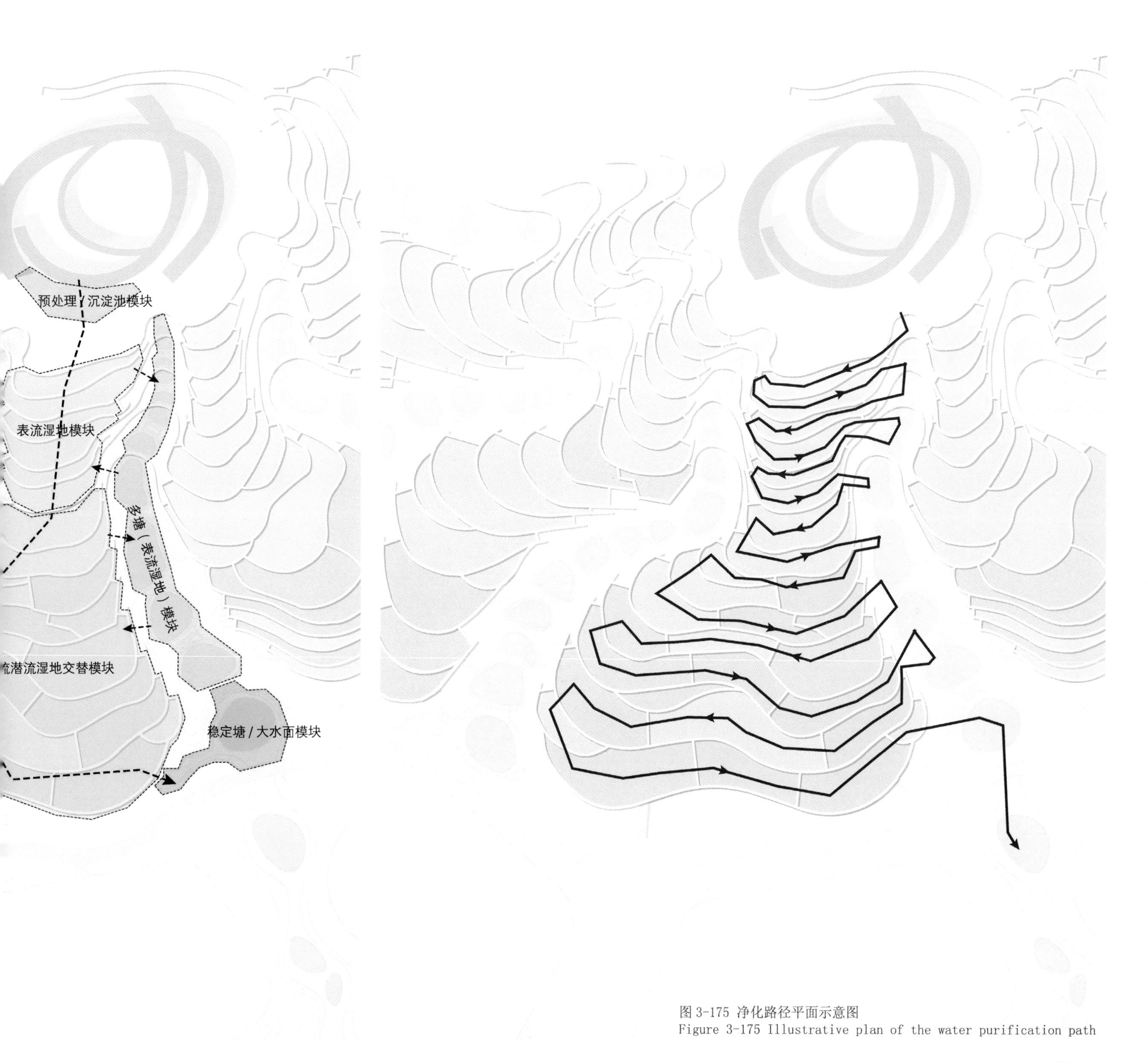

图 3-175 净化路径平面示意图
Figure 3-175 Illustrative plan of the water purification path

（3）生态湿地的植物——美且能净化水质

在清渠如许，植物对湿地净化和景观营造起着重要作用。在湿地中，主要选取了净化能力强、观赏效果好的挺水植物，例如香蒲、黄菖蒲、水生美人蕉、千屈菜、梭鱼草、再力花等，这些植物对悬浮物、氮、磷、COD、BOD 等有很好的去除效果。在岸上，为改良土壤，选取了紫穗槐作为陆生植物基底；同时种植了碧桃、山桃、西府海棠等观赏小乔木，以及金鸡菊、迎春、松果菊、鸢尾等地被植物。

（3）Plants in Wetland—Beautiful and Purify the Water

In Wetland Terraces, plants play an important role in wetland purification and landscape creation. Emergent plants with strong purification ability and good ornamental effects are the main select in the wetlands, such as *Typha orientalis*, *Iris pseudacorus*, *Canna glauca*, *Lythrum salicaria*, *Pontederia cordata*, and *Thalia dealbata* Fraser, etc. These plants have a strong removal effect on suspended solids, nitrogen, phosphorus, COD, BOD, etc. On the lakeshore, in order to improve the soil, *Amorpha fruticosa* Linn was selected as the base of terrestrial plants. Small ornamental trees such as flowering peach, wild peach, and midget crabapple, and ground cover plants such as coreopsis, winter jasmine, echinacea, and iris were planted.

图 3-176 水生美人蕉
Figure 3-176 Aquatic Canna Lily

图3-177 台田种植

Figure 3-177 Terraced planting

（4）生态湿地的制高点——锦绣云台

清渠如许的山顶之上，一座景观环廊宛如双龙以盘旋腾飞之势从绿荫中升起，这便是“锦绣云台”。

（4）Top of Wetland—Cloud Terrace

A landscape corridor rises on the top of Wetland Terraces, which is called the Cloud Terrace.

作为园博会南区的制高点，“锦绣云台”提供了360° 的立体观赏园博园全景的体验，登上云台可俯瞰清渠如许水净化流程并饱览园中其他各区绮丽风光。

At the top pf the southern part of the park, the Cloud Terrace provides a 360° viewing experience of the park. After climbing up the terrace, one can overlook the water purification procedure of Wetland Terraces and enjoy the beautiful scenery of other areas of the park.

图 3-179 锦绣云台顶视图
Figure 3-179 Top view of the Cloud Terrace

图 3-180 锦绣云台观景
Figure 3-180 Viewing platform of the Cloud Terrace

锦绣云台不仅是山顶休憩场所，同时也是一个景观化的净水设施，在特殊模块化表皮肌理之下，隐藏着一套与清渠如许的水净化系统接驳的立体水净化系统，类似于水塔，灌溉植物的富营养水从顶部经过 3 层立体绿化及生态净化基质螺旋式流下，流入梯田水净化系统。

The Cloud Terrace is not only a resting place, but also a landscaped water purification facility. Under the special modular skin texture, there is a set of three-dimensional water purification system that is connected to the water purification system of Wetland Terraces. It works similarly as a water tower, in which the nutrient-rich water for irrigation flows down spirally from the top through the three-levels of greening and the ecological purification system, and is then distributed to the terraced water purification system.

图 3-181 中心水景
Figure 3-181 Central water feature

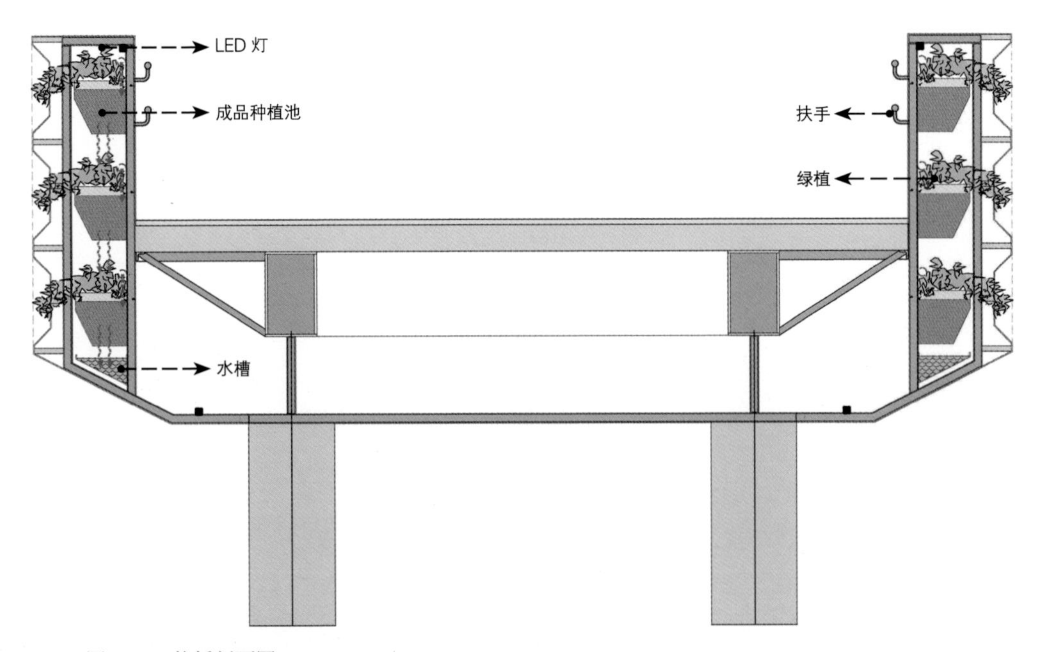

图 3-182 栈桥剖面图
Figure 3-182 Section of skybridge

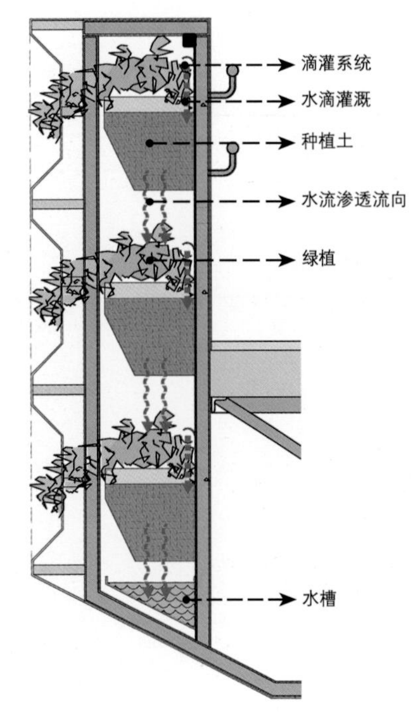

图 3-183 立体净化做法详图
Figure 3-183 Vertical filtration detail

图 3-184 锦绣云台立面
Figure 3-184 Elevation of the Cloud Terrace

图 3-185 清渠如许全景
Figure 3-185 Panorama of the Wetland Terraces

图 3-186 净化台田
Figure 3-186 Terraced filtration field

图 3-187 台田系统
Figure 3-187 Terraced filtration field

图 3-188 梯田栈道
Figure 3-188 Terraced boardwalk

图 3-189 休憩景观盒
Figure 3-189 Lookout pergola

图3-190 栈道架在梯田之上

Figure 3-190 Boardwalks sitting above the terraced filtration fields

3.2.10 景点七 矿坑花园

THE RECLAIMED QUARRY PARK

矿坑花园原为人工采石场，由于长期开采作业形成了一个巨大的矿坑，陡坎随处可见，高差变化很大，且堆满了碎石矿渣、生活和建筑垃圾，土壤受到污染，生态环境破坏严重。

The Reclaimed Quarry Park was formerly an artificial quarry. The long-term mining operations formed a huge pit, in which steep ridges can be seen everywhere, with very large height difference. In addition, the site is piled up with gravel slag, domestic and construction waste with polluted soil—the ecological environment damage was very serious.

图 3-191 矿坑花园改造前实景
Figure 3-191 Photo showing condition of the Reclaimed Quarry Park before the reconstruction

图3-192 矿坑花园鸟瞰图
Figure 3-192 Bird-eye view of the Reclaimed Quarry Park

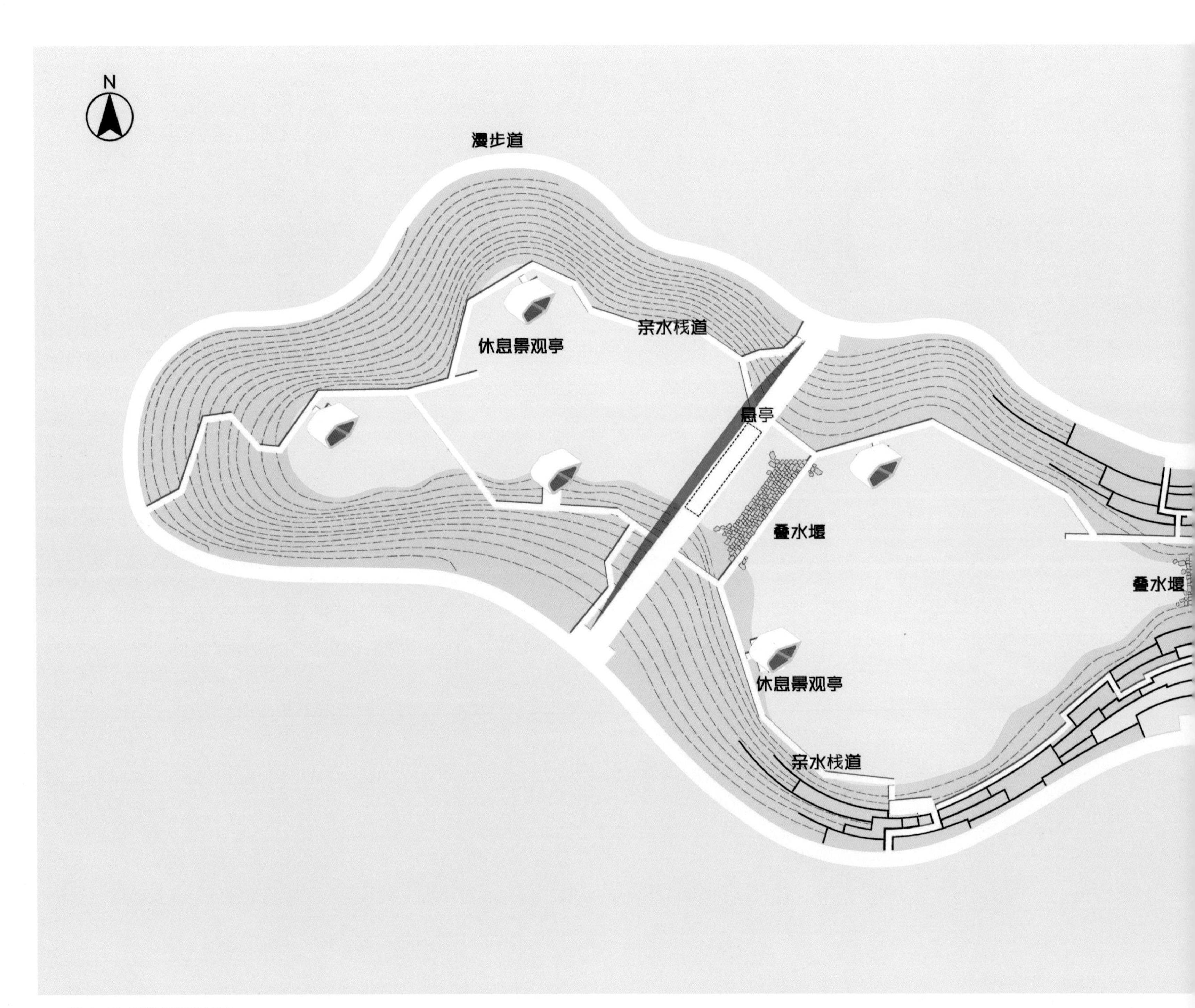

图 3-193 矿坑花园平面图
Figure 3-193 Plan of the Reclaimed Quarry Park

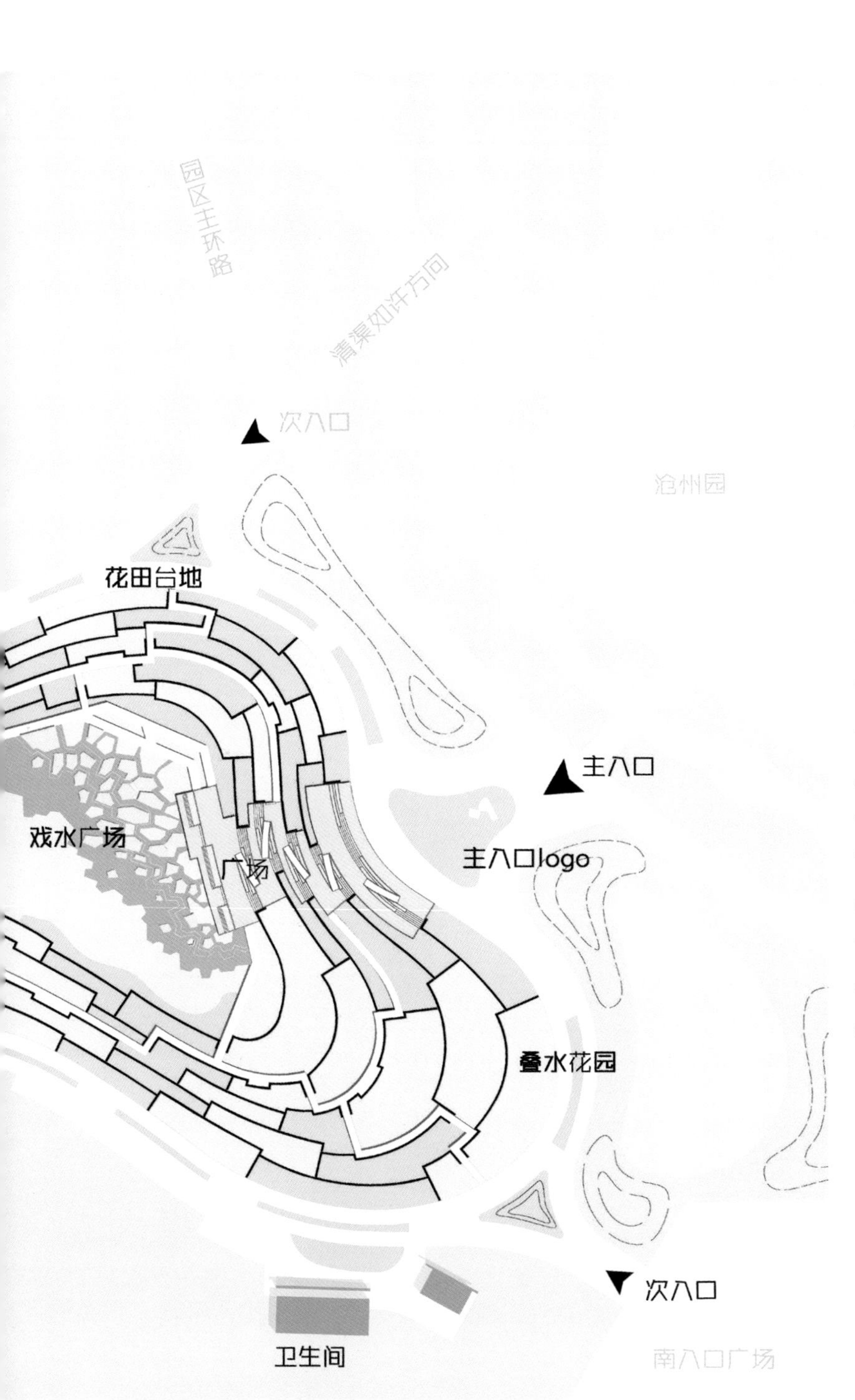

设计特色

DESIGN FEATURES

设计采用三种修复方式综合解决场地土壤污染问题，分别是矿渣填埋、土壤置换和引水净化。将现状矿渣进行填埋封存，部分土壤进行置换，以满足园博会对土质和植被生长的需求，引入了清渠如许湿地净化后的水，结合植物的吸附作用对水体进行深度净化，营造生境多样性，激活场地活力。

设计方案充分利用现状矿坑的坑洼和内凹的地形，形成内向型景观。首先将原有陡坡层叠填土，对部分陡峭区域进行梯田化处理，打造为花田台地。引入清渠如许净化后的水，结合局部台地，形成叠水景观，结合植物的吸附作用进一步净化水体，最终汇流形成三组高差不同的水体景观，分别展现了矿坑修复自然演替的三个阶段。

The design adopts three remediation methods to solve the problem of soil pollution on the site, namely slag landfill, soil replacement and water diversion purification. The current slag is buried and sealed, and part of the soil is replaced to meet the needs of the Expo for soil quality and vegetation growth. The water purified by the Wetland Terraces wetland is introduced, combined with the adsorption of plants, to deeply purify the water body, creating habitat diversity, and activating the vitality of the site.

The design makes full use of the pit terrain of the current quarry and forms an inward-looking landscape. First, the original steep slope is formed into terraced flower fields by earth moving. The purified water from Wetland Terraces is introduced to form cascading water landscape, making full use of the adsorption effect of plants to further purify the water body, and finally converge the water to form three groups of water landscapes of different heights, which respectively demonstrate the three stages of natural succession of the quarry restoration.

第一个阶段是基底恢复，表现的是对原旧迹的破坏、新生命的开始，由龟裂的混凝土石板组成，自北向西混凝土板的缝隙逐渐扩大，形成硬质向自然的过渡，点状种植，丰富视觉。

The first stage is the restoration of the foundation, which represents the destruction of the original and the beginning of a new life. It is composed of cracked concrete slabs, and the gaps between the concrete slabs gradually expand from north to west, forming a transition from rigidity to nature. Dot planting is introduced here to enrich the vision.

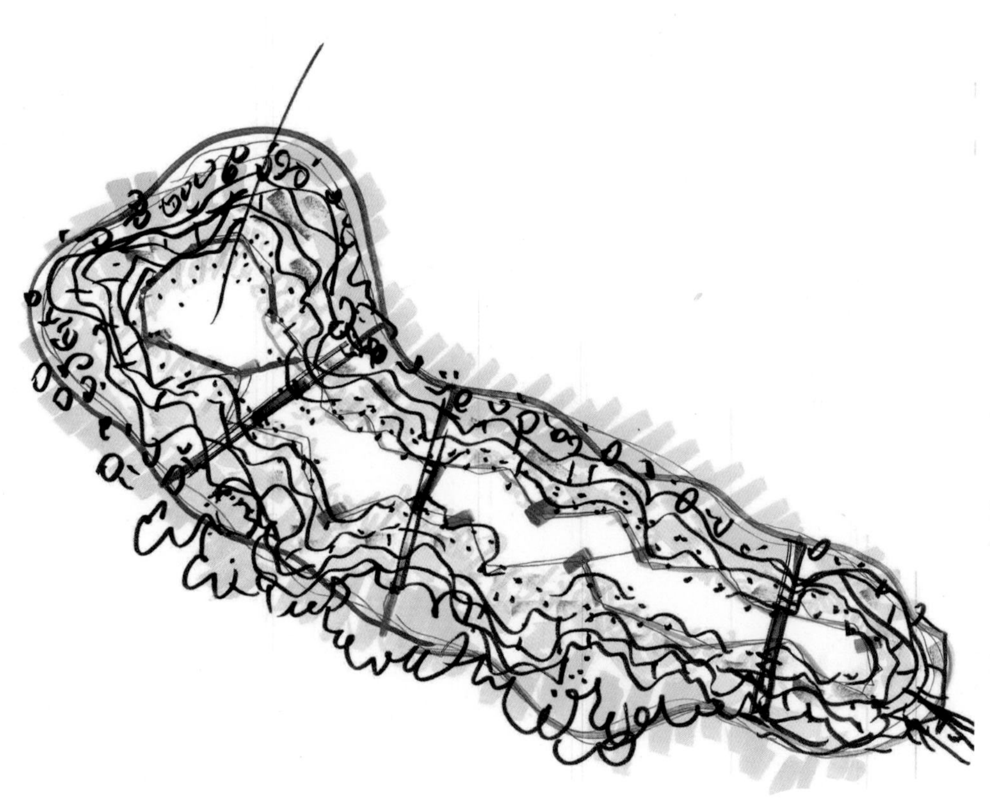

图 3-194 首席设计师俞孔坚矿坑花园草图
Figure 3-194 Sketch of the Reclaimed Quarry Park by chief designer Yu Kongjian

图 3-195 入口广场与趣味水景鸟瞰图
Figure 3-195 Bird-eye view of the entry plaza and playful water feature

图 3-196 趣味水景一
Figure 3-196 Playful water feature (I)

图 3-197 趣味水景二
Figure 3-197 Playful water feature (II)

5G+

图 3-198 矿坑花园生态环境修复
Figure 3-198 Ecological habitation of the Reclaimed Quarry Park

图 3-199 景观亭
Figure 3-199 Outlook pergola

第二个阶段展现的是植被修复。豁然开朗的视线，丰富多彩的植物，旧址逐渐消失，生态基底逐步建立。

The second stage showcases vegetation restoration. The sudden clear sight, the colorful plants, the gradually disappearing relics, all together represent a gradually established ecological base.

图 3-200 第二阶段：植被修复
Figure 3-200 Phase two: vegetation restoration

沿着步道继续往前，到达第三个阶段，展示的是矿坑修复的最终阶段——生境修复，两侧台田逐渐消失，自然缓坡入水，充满了自然野趣。

Walking along the boardwalk, visitors will reach the third stage, which shows the final stage of the quarry pit restoration—habitat restoration. The terraces on both sides gradually disappear, replaced by natural gentle slopes to enter the water.

图 3-201 第三阶段：生境修复
Figure 3-201 Phase three: ecological environment restoration

悬亭是整个矿坑花园的点睛之笔，位于长 85m 的人行桥下方。悬亭的灵感来自于传统园林中的“水榭”，全景玻璃盒悬挂于人行栈桥之下，长 23m，高 3m，由钢结构搭建框架，外包钢化玻璃，是景区内的视觉中心，也是游人欣赏湖景的最佳场所。

The hanging pavilion is the finish touch of Reclaimed Quarry Park. Located under the 85m long footbridge, the hanging pavilion is inspired by the “water pavilion” in Chinese traditional garden. The panoramic glass box is suspended under the pedestrian trestle. It is 23m long and 3m high, framed by a steel structure and covered with tempered glass. It is the visual center of the park, thus the best place for visitors to appreciate the lake.

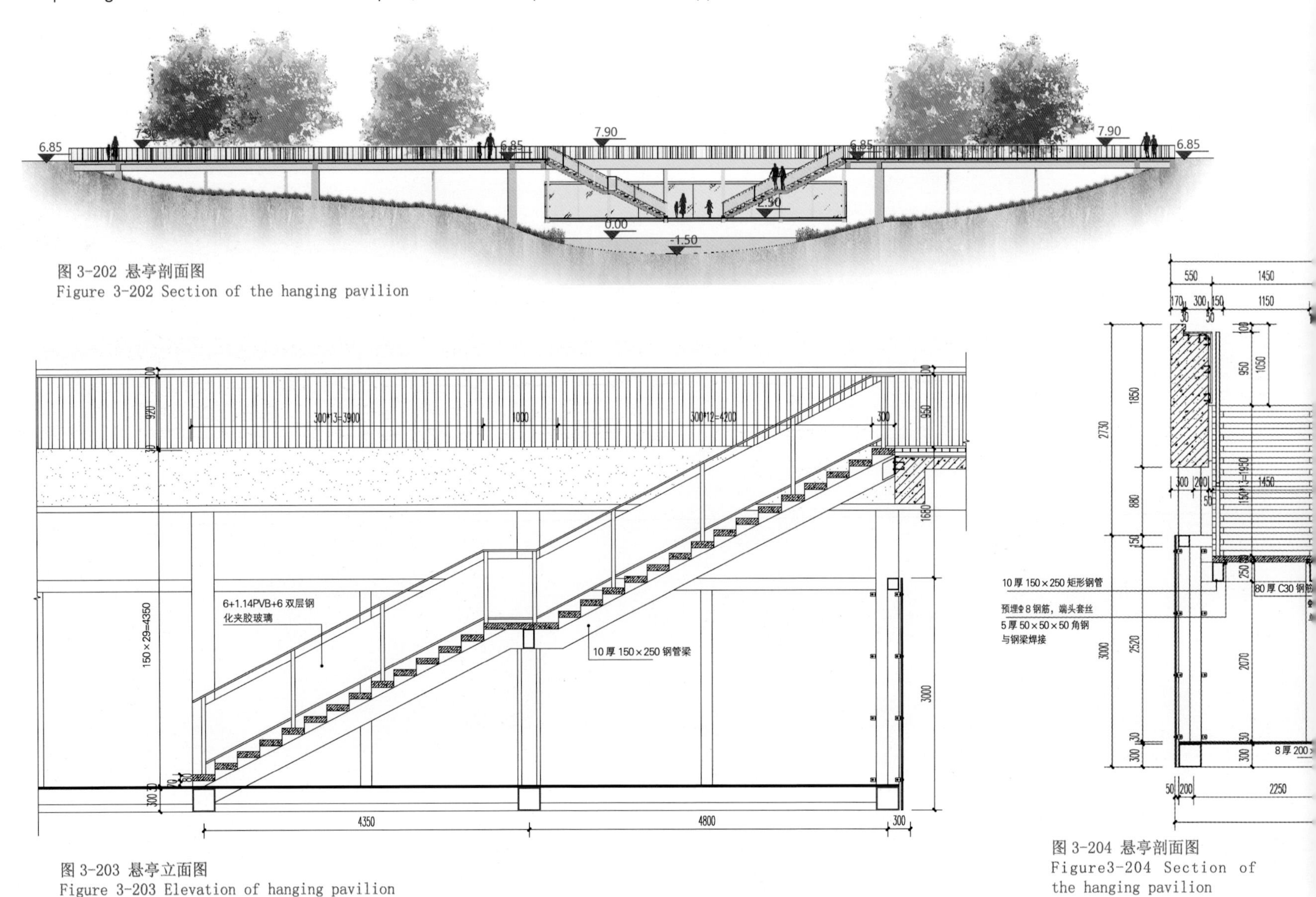

图 3-202 悬亭剖面图
Figure 3-202 Section of the hanging pavilion

图 3-203 悬亭立面图
Figure 3-203 Elevation of hanging pavilion

图 3-204 悬亭剖面图
Figure3-204 Section of the hanging pavilion

图 3-205 悬亭入口
Figure 3-205 Entrance of the hanging pavilion

图 3-206 悬亭
Figure 3-206 Hanging pavilion

图 3-207 草花组合
Figure 3-207 Flower mix

矿坑花园以生态修复为理念，运用多种景观营造手法，成为生态修复主题和景观设计相结合的典范。

矿坑花园种植以“叠花立木，鸢尾写幽”为主题，跌水区的水生植物群落、多彩花带与葱郁的观赏草同时结合现状乔木树群，新增楸树林与多种草花及湿生水生植物，兼顾水质净化功能，形成现代复合型群落结构。

Reclaimed Quarry Park's design is based on the concept of ecological restoration. It uses a variety of landscape construction techniques to make it a model of the combination of ecological restoration and landscape design.

The planting design uses stacked flowers, standing trees and iris to create a quiet atmosphere. The aquatic plant community, colorful flower belts and lush ornamental grasses in the cascade area are combined with the existing arbor tree group. Added with catalpa forest and a variety of flowers and aquatic plants, it forms a modern complex community structure that takes into account the water purification function.

图 3-208 跌水区
Figure 3-208 Cascade area

3.2.11 景点八 梦泽飞虹

THE FLYING BRIDGE

“梦泽飞虹”是一座形态优雅的白色栈桥。桥体采用流动线条设计手法，像一条彩虹在湖面和栖岛之间穿过，成为西湖水库北侧的视觉焦点，并作为连接西湖水库的主要纽带。

The Flying Bridge is an elegant white trestle bridge. The bridge body looks like a rainbow passing between the lake and the habitat. While it performs as the link connecting the Xihu Dam, it becomes the visual focus on the north side of the Xihu Dam.

图 3-209 梦泽飞虹鸟瞰图
Figure 3-209 Bird-eye view of the Flying Bridge

由于园区的中部被沁河、西湖水库、溢洪道隔开，因此修建一条全长约1.5km、从水库北侧边缘“划过”的多功能栈桥，使得园区内部可以形成一条完整的主要交通立体环线。

Since the center part of the park is separated by the Qin River, the Xihu Dam, and the spillway, building a multifunctional trestle bridge with a total length of 1.5 kilometers that “sweeps” through the northern edge of the reservoir helps form a main three-dimensional traffic loop inside the park.

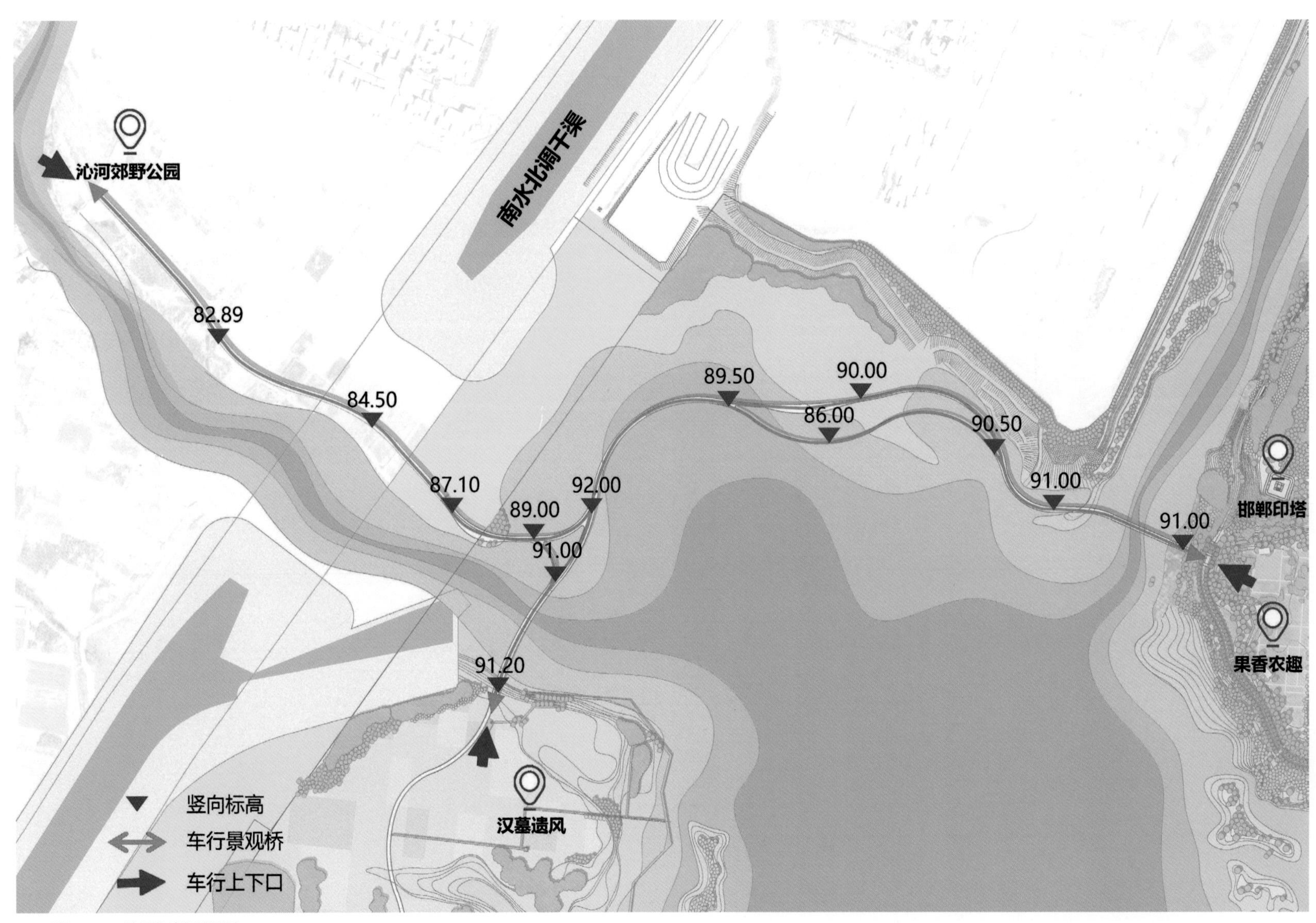

图 3-210 梦泽虹桥平面图
Figure 3-210 Plan of the Flying Bridge

设计特色
DESIGN FEATURES

梦泽虹桥的宽度需要满足步行交通和电瓶车通行的要求。设计采用了平行、交叉式的桥上空间模式：在3个主入口处加宽桥面，满足步行和车行同时上桥之后，随即将步行空间拆分出来，依桥的形态和周边景观的变化设计两层立体桥面，高可远眺湖景，低可观鸟亲水，使整个桥上漫步变为一次独特、多样的空中体验。

The width of the Flying Bridge needs to meet the requirements of both the pedestrian traffic and cart traffic. The design adopts a parallel and intersecting space model on the bridge: widen the bridge deck at the three main entrances to allow pedestrians and vehicles on the bridge at the same time. Then split the pedestrian space, making a two-level bridge according to the surrounding landscape. Visitors can overlook the lake on the high spots, and watch the birds and appreciate the water while on low spots. This makes the walk on the bridge a unique and diverse aerial experience.

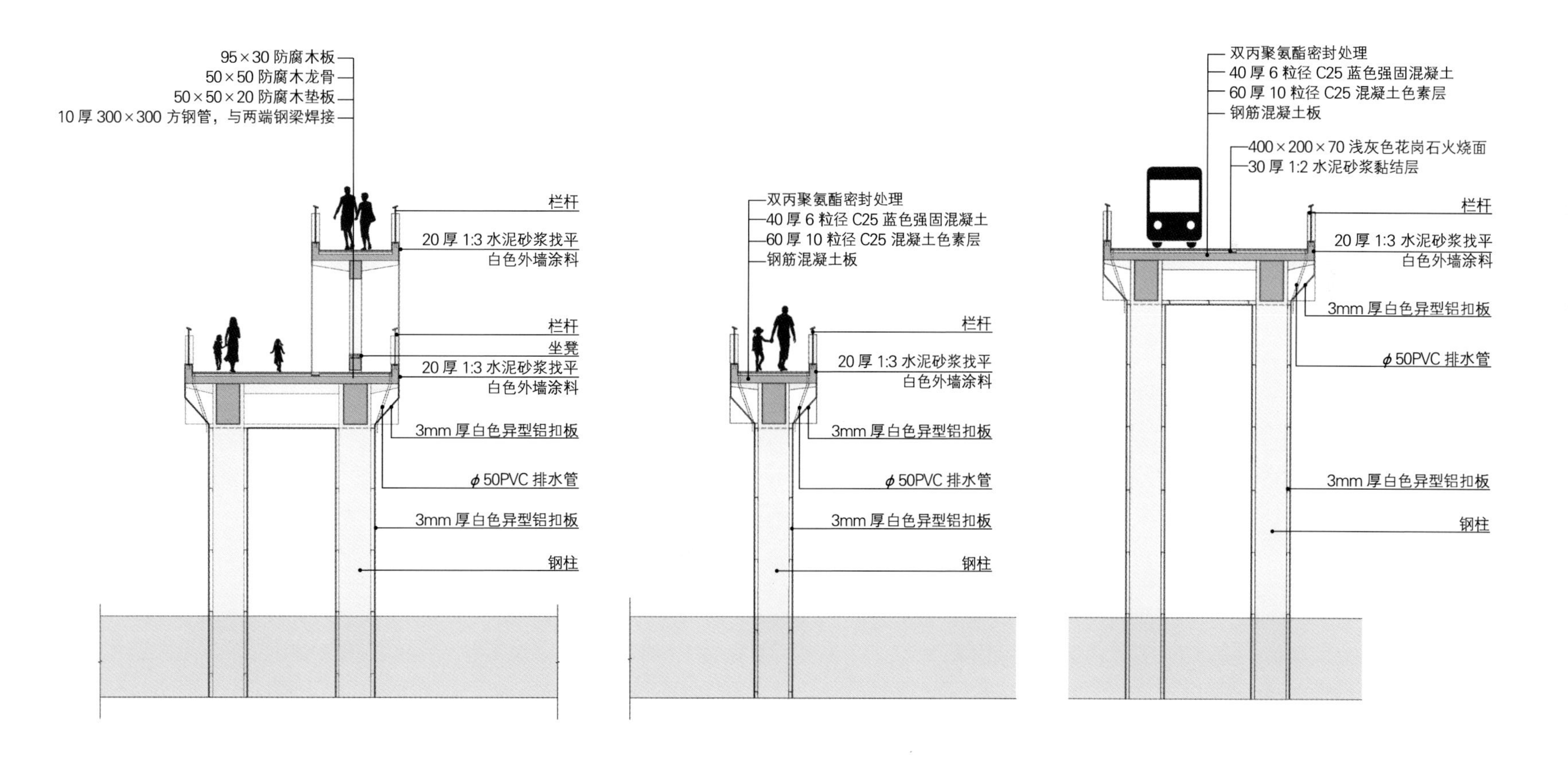

图 3-211 梦泽虹桥剖面图
Figure 3-211 Section of the Flying Bridge

桥体的立面设计根据桥形的走向以及外挂铝板和栏杆的渐变呈现起伏，在保持整体感的前提下，让桥的侧面从视觉上显得更加轻薄，营造出桥体在水上漂浮的灵动感，再通过灯光变化勾勒出桥身的婀娜，让夜间的梦泽虹桥更加虚幻动人。

The facade design of the bridge body shows ups and downs according to the direction of the bridge shape and the gradual change of the external aluminum panels and railings. On the premise of maintaining a sense of integrity, the side of the bridge is made visually lighter and thinner, creating an impression of the bridge body floating on the water. The change of light draws the gracefulness of the bridge outline, making the Flying Bridge even more illusory and moving at night.

图 3-212 梦泽虹桥立面
Figure 3-212 Elevation of the Flying Bridge

图 3-213 梦泽虹桥全景一
Figure 3-213 Panorama of the Flying Bridge (I)

图 3-214 梦泽虹桥全景二
Figure 3-214 Panorama of the Flying Bridge (II)

图3-215 梦泽虹桥局部（跨芳草寻鹤）
Figure 3-215 Partial view of the Flying Bridge (passing through the wetland)

图 3-216 梦泽虹桥局部（跨溢洪道）
Figure 3-216 Partial view of the Flying Bridge (across from the overflow canal)

3.2.12 景点九 印塔夕照

HANDAN TOWER IN THE SUNSET

邯郸印塔作为邯郸园博园中的制高点，总共分为 7 层，塔高 27.6m。

每当夕阳西下，塔影横空而立，便可远望芳草寻鹤，波光辉映，景色至为美胜，故被称为“印塔夕照”。

As the commanding height of the whole park, Handan Tower is 27.6 meters high with 7 floors.

When the sun goes down, the shadow of the tower stands up, visitors can look at the grass from distance and look for cranes. This beautiful scenery was named Handan Tower in the Sunset.

图 3-217 邯郸印塔一
Figure 3-217 Handan Tower (I)

图 3-218 印塔夕照
Figure 3-218 Handan Tower in the Sunset

设计特色

DESIGN FEATURES

设计力求用创新的手法体现邯郸的城市精神与文化内涵。方案通过“雕刻”的概念在塔身上展现出虚实相间的肌理，使其像一枚传统印章一般矗立在园博园的山水之间。塔身立面呼应小篆中“邯郸”二字的基本笔画，展现出场地的文化内涵。塔心以红色呈现，与浅灰色的外立面强烈对比，增加了印塔的视觉冲击力，配合夜景灯光更是塑造出印塔通透、鲜明的形象，成为园博园又一标志性的景观符号。

The design uses innovative methods to reflect the urban spirit and cultural connotation of Handan. The texture of combining the virtual and the real is showed on the tower facade by “carving”, making the tower stand like a traditional seal. The facade of the tower echoes the basic strokes of the word “Handan” in the Small Seal style (calligraphy). The center of the tower is presented in red, which is in sharp contrast with the light gray facade, increasing the visual impact of the tower. In conjunction with the night scene and lighting, it creates a transparent and clear image of the tower.

篆书始创于秦，笔法瘦劲挺拔，曲线较多，直线较少，还保存着古代象形文字的显著特点。

景观塔设计创意来自于篆书“邯郸”二字，分解笔画，设计成镂空写在塔体上的效果，体现出历史名城邯郸深厚的文化城市底蕴。

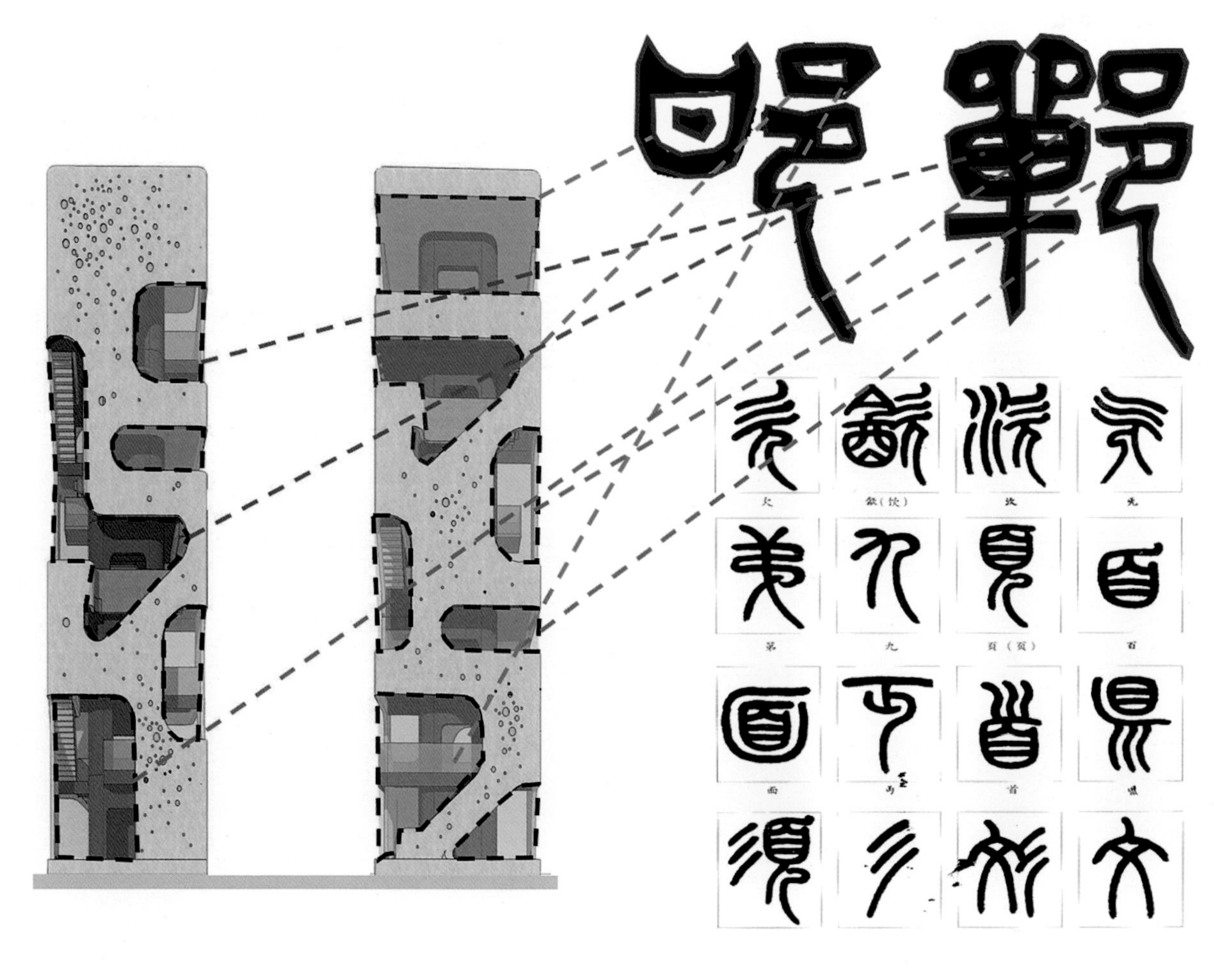

图 3-219 文化内涵
Figure 3-219 Cultural connotation

图 3-220 邯郸印塔二
Figure 3-220 Handan Tower (II)

图 3-221 邯郸印塔施工过程
Figure 3-221 Construction process of the Handan Tower

为了使设计理念落地，达到浑然天成的艺术效果，印塔外立面的处理至关重要。通过对各种材料和可能性的综合判断，设计最终选择了预制混凝土挂板作为外饰面。挂板材料采用可回收废料二次加工而成，在节约成本的同时体现了环保低碳的建筑理念。挂板在异地精确预制，通过工厂设备将电脑中细致推敲的 3D 曲线形体准确落实，再运输至现场进行组合安装，整个过程犹如完成一幅立体无缝拼图一般精准。

In order to make the design concept come with organically artistic effect, the facade design is of great significance. After studying various materials and possibilities, the design team chose precast concrete as the exterior material. The claddings of which are made by recycled waste, which embraces the concept of environmental protection and low-carbon architecture as well as cost efficiency. The concrete panels were prefabricated in the factory, and then transported to the site for assembly and installation. The whole process was seamless and accurate.

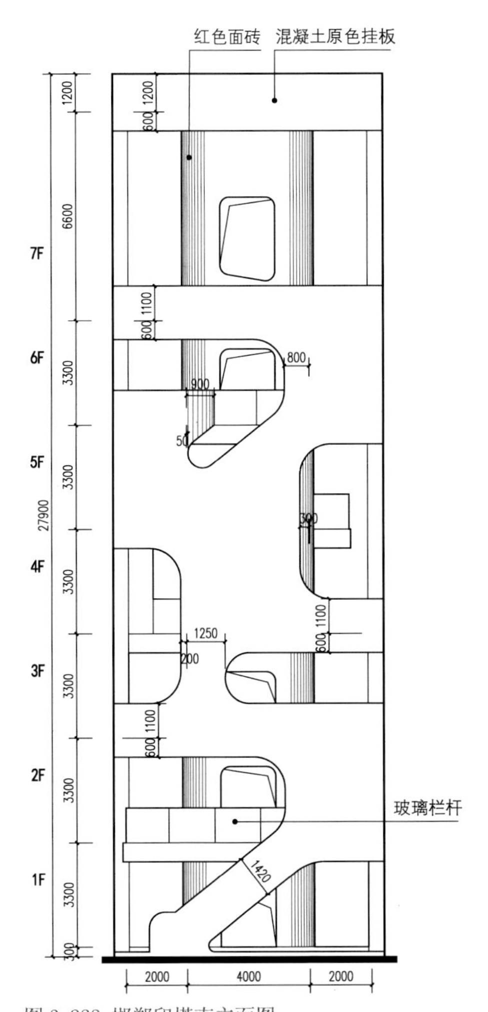

图 3-222 邯郸印塔南立面图
Figure 3-222 South elevation of the Handan Tower

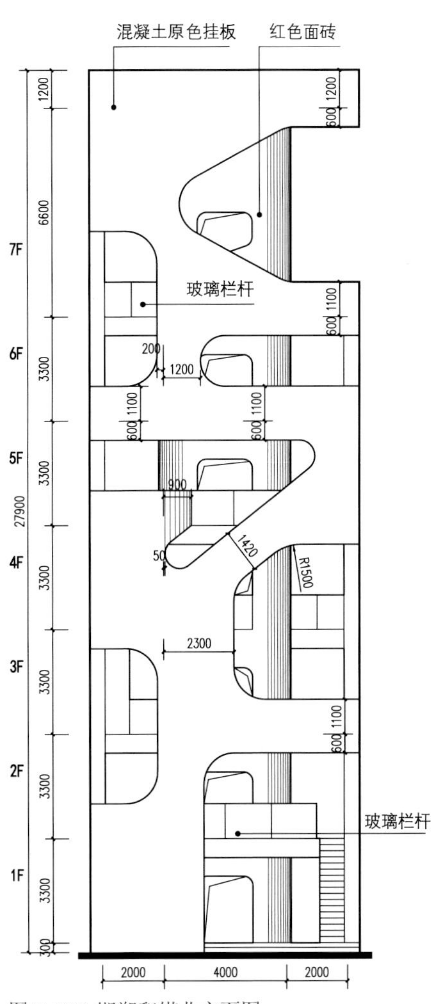

图 3-223 邯郸印塔北立面图
Figure 3-223 North elevation of the Handan Tower

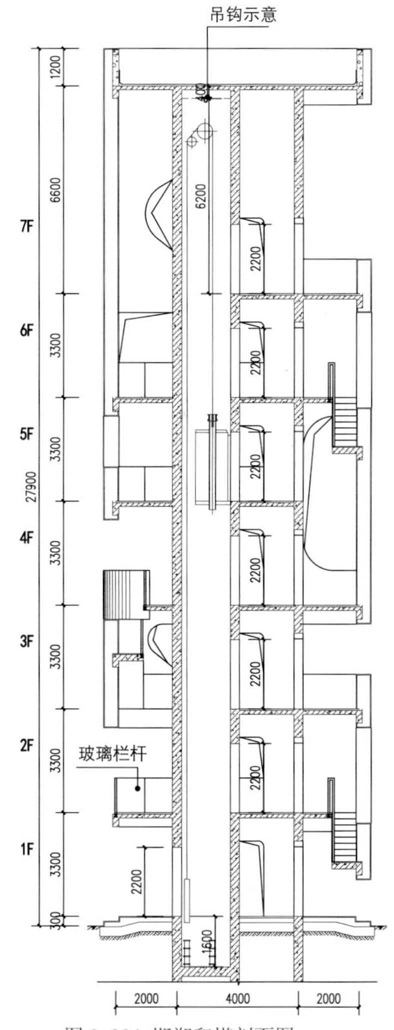

图 3-224 邯郸印塔剖面图
Figure 3-224 Section of the Handan Tower

图 3-225 夜幕下的邯郸印塔剪影
Figure 3-225 Silhouette of Handan Tower in the dusk

邯郸印塔虽然是一个尺度相对小的景观建筑，但在整体结构上借鉴了高层建筑的设计原理，即以一个塔身“核心筒”为基础，支撑起各层楼板和立面挂板。核心筒内部设置有一部电梯，方便游人直接登塔的同时，缩小塔身截面面积，使其从视觉比例和承重体量上都能满足设计要求。这样的设计可将塔中的停留空间保持在外侧，让登塔的人能在每一层都获得 360° 无阻碍观景体验。

Although Handan Tower is a relatively small-scale architecture, the overall design principle is derived from the high-rise buildings in its structure. Using the tower's central tube as the core, the floors and facade siding are propped up level by level. An elevator is installed inside the core tube to allow visitors to directly reach the top, while reducing the sectional area of the tower body for both visual proportions and load-bearing volume considerations. The staying space in the tower is arranged outside of the core tube, for visitors to enjoy a 360° viewing experience on each floor.

图 3-226 360° 观景体验
Figure 3-226 360° viewing experience

图 3-227 远望邯郸印塔
Figure 3-227 Looking Handan Tower from distance

图3-228 掩映花海后的邯郸园塔
Figure 3-228 Handan Tower shading in the back of the flower field

3.2.13 景点十 民俗印象

FOLK CULTURE PAVILION

民俗印象以邯郸民间非物质文化遗产为主题，利用现代景观的表现手法具象化，使参观者能够在不断变化的景观空间中亲身触碰到民俗文化的独特魅力，充分调动人的各种感官，为游人带来一场沉浸式的文化景观秀。

The Folk Culture Pavilion takes Handan's folk intangible cultural heritage as the theme, and uses modern landscape expression techniques to visualize it. Visitors can personally touch the unique charm of the folk culture, with their various senses fully mobilized.

图 3-229 民俗印象平面图
Figure 3-229 Plan of the Folk Cultural Pavilion

图 3-230 民俗印象鸟瞰图
Figure 3-230 Bird-eye view of the Folk Cultural Pavilion

展园特色详解
GARDEN DETAILS

（1）逐梦园

梦文化为邯郸十大文化脉系之一，有著名的“黄粱一梦”典故，在全国也具有唯一性，因此邯郸也被称为“美梦之乡”。逐梦园是民俗印象系列的第一个景观单元，园中通过空间尺度和镜面材料的变化，向游人展示梦的五个阶段——波动、专注、平静、探索和光明。

（1）Dream Pavilion

Dream culture is one of the ten major cultural lines in Handan. The famous Chinese literary allusion “A Golden Millet Dream” is originated in Handan. Therefore, Handan is also known as the “hometown of beautiful dreams”. Dream Pavilion is the first landscape unit in the series Folk Culture Pavilions. Through the changes of spatial scale and mirror materials, the garden shows visitors the five stages of dreams—fluctuation, concentration, tranquility, exploration and luminosity.

图 3-231 逐梦园鸟瞰图
Figure 3-231 Bird-eye view of the Dream Pavilion

梦的五个阶段：

第一阶段“波动”，通过一条长长的嵌入式甬道，利用地形和镜面不锈钢板的组合来控制空间感受和光线，将游客从现实带入梦境世界，在光线昏暗、狭窄的空间里，游客的心里会有一种波动不安感。

第二阶段“专注”，通过控制游人视线，使其在黑暗世界中将注意力集中在自我之上。

第三阶段“平静”，带着游人突破狭小的空间，来到一处开阔境地，在过程中重新静下心来，准备好进入更深层次的梦境。

第四阶段“探索”，带领游人穿过水面到达园区主空间中央，在水雾中尽力分辨真实的世界与水中的倒影。

第五阶段“光明”，透过光线让游人在镜面不锈钢柱阵中邂逅无数个“自己”，在梦醒时刻的亦真亦幻中完成逐梦园游览体验。

Five Stages of Dreams:

Stage One “fluctuation” : through a long embedded corridor, using the combination of the terrain and the reflective stainless steel to control the space experience and light, the garden brings visitors from reality into the dream world. In a dimly lit and narrow space, the visitor will have a sense of volatility in his heart.

Stage Two “concentration” : by controlling the sight of visitors, the garden makes them focus on themselves in the dark world.

Stage Three “tranquility” : takes visitors to break through the narrow space and come to an open place. Visitors calm down again and prepare to enter a deeper dream.

Stage Four “exploration” : leads visitors through the water to reach the center of the garden, making them try to distinguish the real world from the reflection in the water in the mist.

Stage Five “luminosity” : allows visitors to encounter countless “selves” in the reflective stainless steel pillar array, and complete the experience of visiting the dream in the real and illusory moments of awakening.

图 3-232 不同镜面的组合
Figure 3-232 Combination of different mirrors

图 3-233 第三阶段：平静
Figure 3-233 Phase three: tranquility

图 3-234 第四阶段：探索
Figure 3-234 Phase four: exploration

图 3-235 第五阶段：光明
Figure 3-235 Phase five: luminosity

图 3-236 镜面不锈钢柱
Figure 3-236 Reflective stainless steel pillars

（2）醉香园

醉香园展现的是邯郸非物质文化遗产之一——贞元增酒传统酿造工艺。设计以其酿造工艺为主题，隐喻利用大地智慧的古老农业文化历史。整个区域占地约 2200m²，由椭圆螺旋步道、种植台地和一个广场组成。

（2）Mellow Wine Garden

Mellow Wine Garden displays one of Handan's intangible cultural heritage—the traditional brewing process. Taking the brewing process as the theme, metaphorizing the history of ancient agricultural culture using the wisdom of the earth. The entire garden covers an area of about 2,200 square meters, consisting of an elliptical spiral trail, planting terraces and a square.

图 3-237 醉香园鸟瞰图
Figure 3-237 Bird eye view of the Mellow Wine Garden

游客由圆形广场进入，沿半下沉式步道缓步下降进入园景内部，步道外侧有一处文化展示墙，由锈红耐候钢板制成，上面有镂空雕刻的展示邯郸酒文化的酿酒工艺图样和工艺介绍文字。步道内侧为特色种植池，在高粱、美人蕉的包围下，沿着步道寻着酒香来到广场当中。广场中央有三个圆环状喷雾装置带，喷出带着酒香的水雾，让游人仿佛置身于酿酒现场之中。

Visitors enter from the circular square and descend slowly along the semi–sunken trail into the garden. There is a cultural display wall made of rust–resistant weathering steel plate, on which a hollow carved pattern shows the winemaking process and craft introduction text of Handan wine. Inside the trail there is a special planting pool. Surrounded by sorghum and canna, visitors walk along the trail to the square to find the aroma of wine. There are three circular spray device belts in the center of the square, spraying out water mist with the wine aroma.

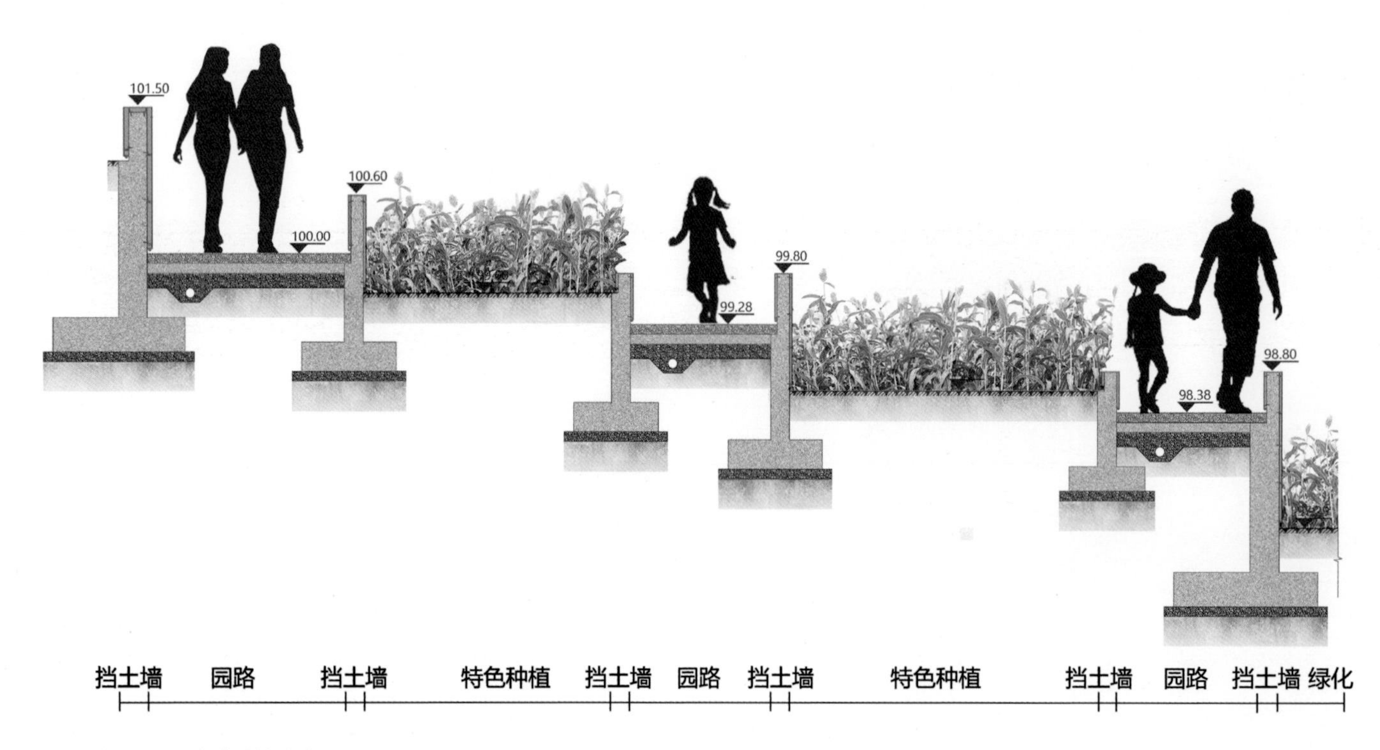

图 3-238 醉香园剖面图
Figure 3-238 Section of the Mellow Wine Garden

图 3-239 特色种植池
Figure 3-239 Customised planters

（3）结草园

邯郸魏县土纺土织及大名县的大名草编被列入 2008 年第二批国家非物质文化遗产名录。结草园以“大名草编”为设计主题，以“编织”为设计理念，将传统工艺用现代景观的语言重新诠释，将传承千年的技艺和现代工艺融入这片场地之中。

设计以一条“草编”和“藤编”交织的廊架为主体，上面有方形的不规则开洞，用抽象的手法表现出传统技艺的独特魅力。

（3）Straw Weaving Garden

The native spinning and weavings of Wei County, and Daming Straw Weaving of Daming County were included in the second batch of National Intangible Cultural Heritage List in 2008. The Straw Weaving Garden draws inspiration from Daming Straw Weaving. It reinterprets traditional craftsmanship in the language of modern landscape, and integrates thousand–year–old skills and modern craftsmanship into the site.

The design is composed of a “straw–woven” and “rattan–woven” pergola as the main body, with irregular square openings on the top, showing the unique charm of traditional techniques with abstract design.

图 3-240 结草园鸟瞰图
Figure 3-240 Bird-eye view of the Straw Weaving Garden

值得一提的是展园中的坐凳采用同样的编织形式，与构筑物相呼应，并且结合步行小径分区域矗立不同颜色的竹子，游人在廊下游览或小憩的过程中就可以欣赏到邯郸本地编织中的特色纹理，由此感受传统工艺带来的震撼。

廊架的局部顶棚上还点缀手工编织小装置，增添区域的趣味性。

The benches in the garden adopt the same woven form to echo the structure, and combined with the walking path, bamboos of different colors are set upright in the area. Visitors can appreciate the special texture of Handan's local weaving while touring or taking a rest under the corridor.

Part of the ceiling of the pergola frame is also dotted with small hand-woven devices to add fun to the area.

图 3-241 不同颜色的竹子
Figure 3-241 Colourful bamboo sticks

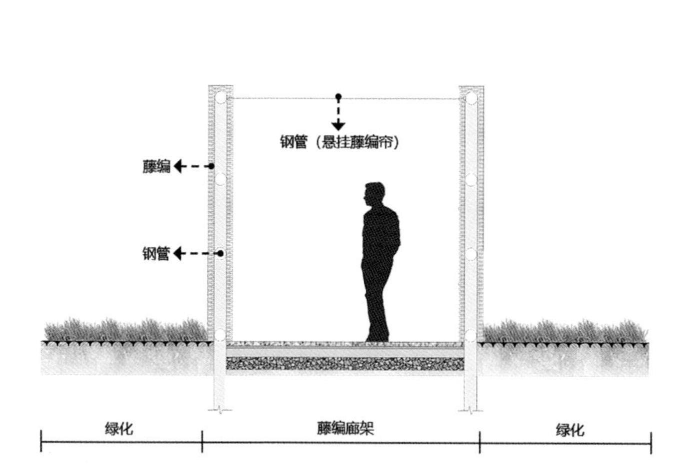

图 3-242 藤编廊架剖面图一
Figure 3-242 Sectional view of the bamboo corridor (I)

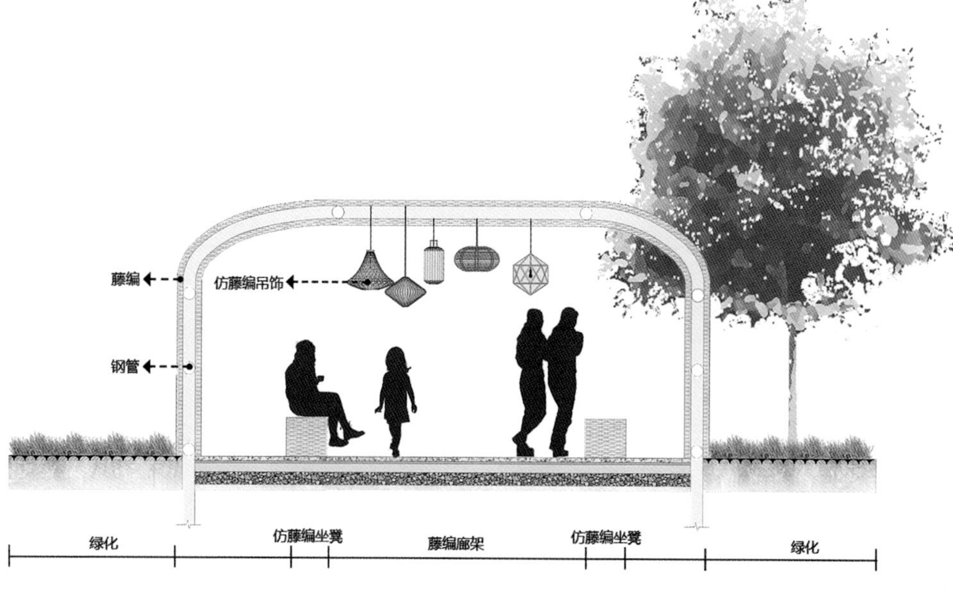

图 3-243 藤编廊架剖面图二
Figure 3-243 Sectional view of the bamboo corridor (II)

图 3-244 悬挂藤编帘
Figure 3-244 Hanging rattan-weaved curtains

图 3-245 悬挂仿藤编吊饰
Figure 3-245 Hanging replica of rattan-weaved deco

（4）诗词园

邯郸是全国首屈一指的成语典故之都，与邯郸市有关的汉语成语典故达1500多个。

诗词园是民俗印象最大的一个展园，也是位于南侧的最后一个园子，展现的是邯郸千年的成语文化。

设计以战国至魏晋时期的书写材料“竹简”为意向，两条曲折步道串联起12面青砖墙，使得整个园区如竹简书卷一般缓缓展开，让人在舒缓的节奏中感受古诗词的韵律灵感。

（4）Poetry Garden

Handan is the nation's leading city of idiom and allusion. There are more than 1,500 Chinese idioms and allusions related to Handan.

Representing Handan's thousand-year-old idiom culture, the Poetry Garden is the largest garden within the Folk Culture Pavilion, as well as the last garden on the south part.

The design is based on the writing material “Bamboo Slips” from the Warring States Period to the Wei and Jin Dynasties. Two zigzag trails are connected with 12 black brick walls, making the entire park slowly unfold like a bamboo slips scroll.

图3-246 诗词园鸟瞰图
Figure 3-246 Bird eye view of the Poetry Garden

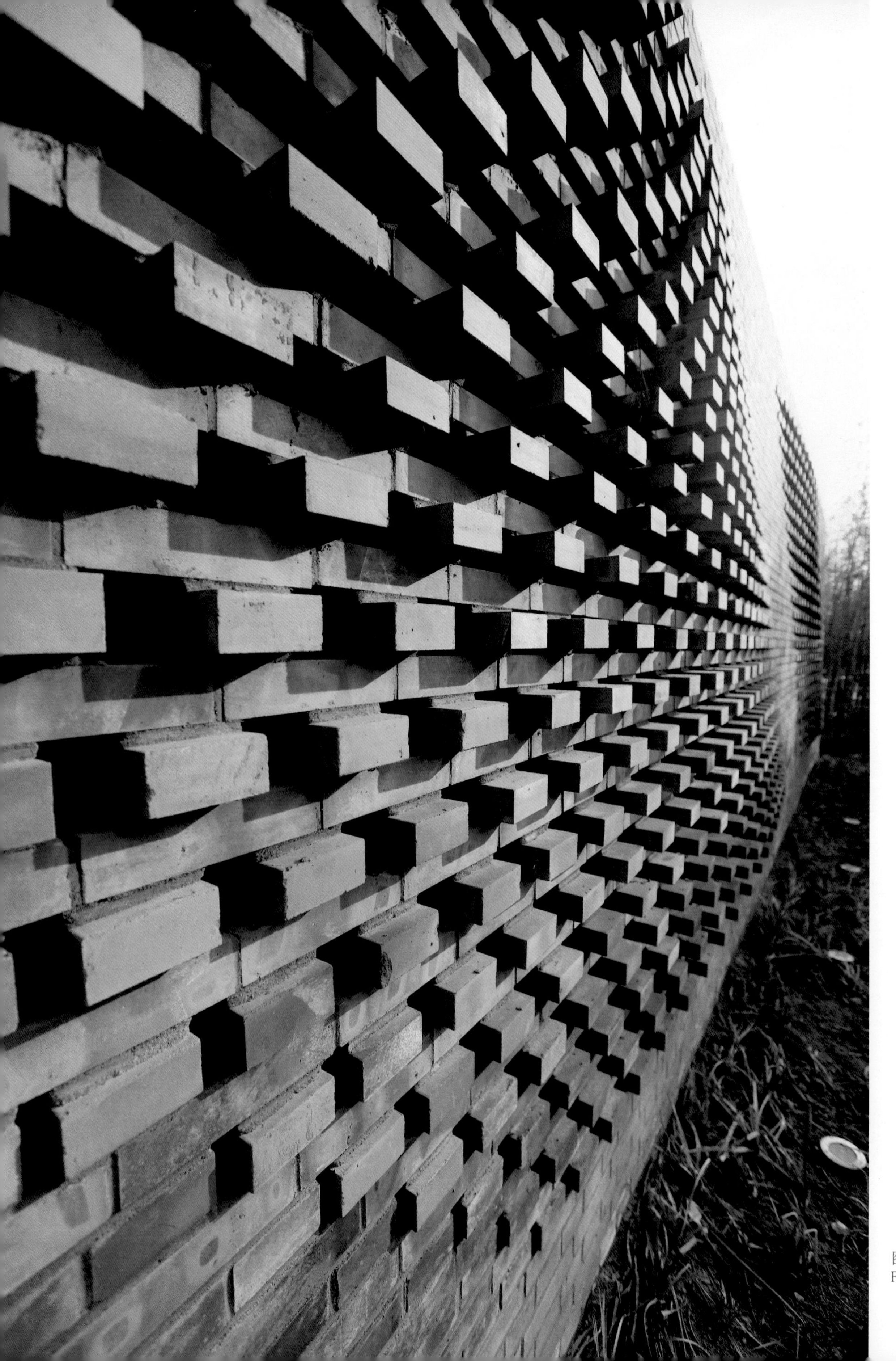

图 3-247 青砖墙
Figure 3-247 Black brick wall

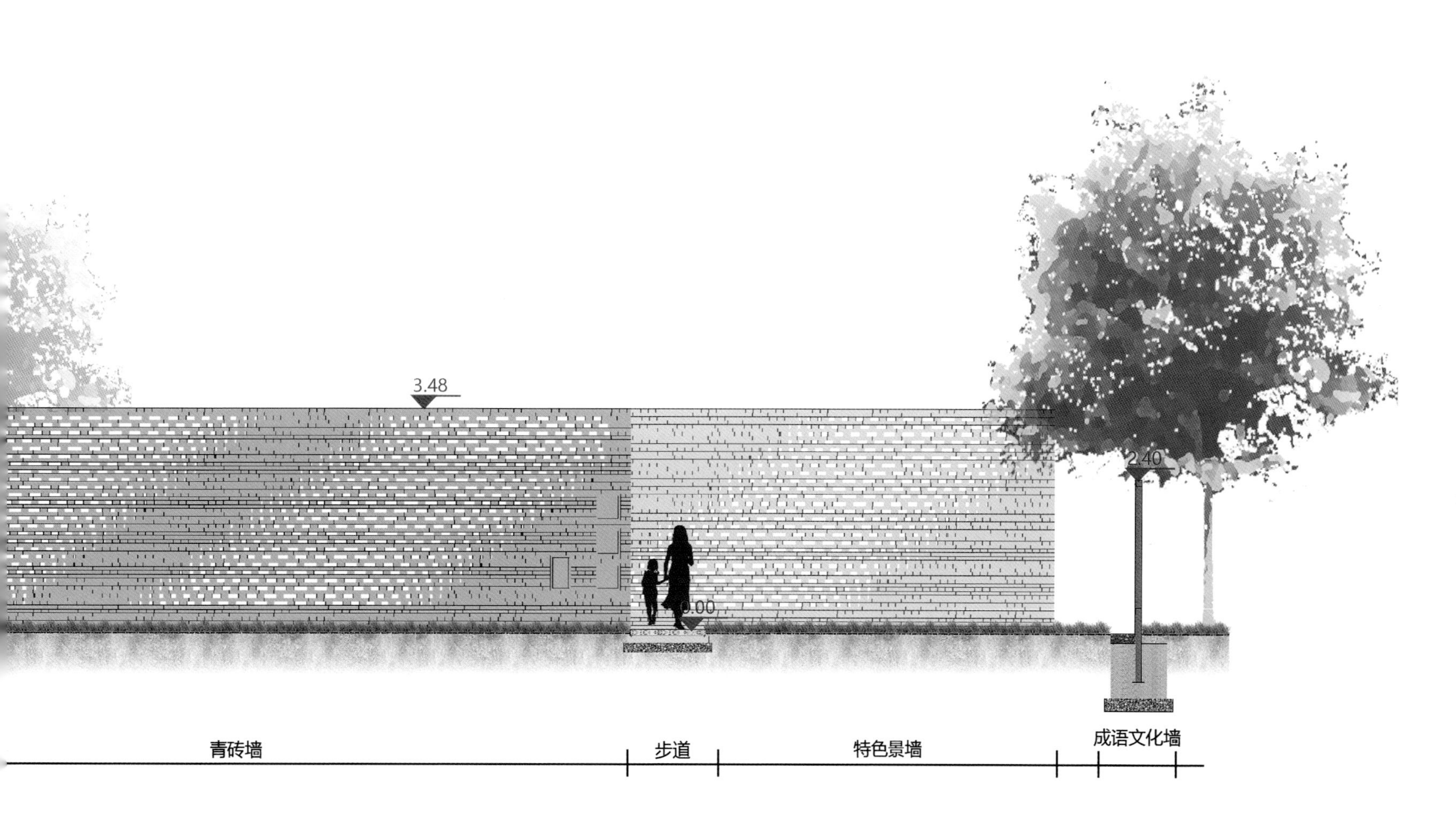

图 3-248 诗词园剖面图
Figure 3-248 Section of the Poetry Garden

园区东侧则是一整面成语装置墙，游人可以在这里找到160多条与邯郸直接相关的成语。在夜晚成语墙还会被灯光点亮，为游人带来更加惊奇的视觉效果。

An idiom wall is installed on the east side of the garden, where visitors can find more than 160 idioms directly related to Handan. At night, the idiom wall lit by lights brings amazing visual effect.

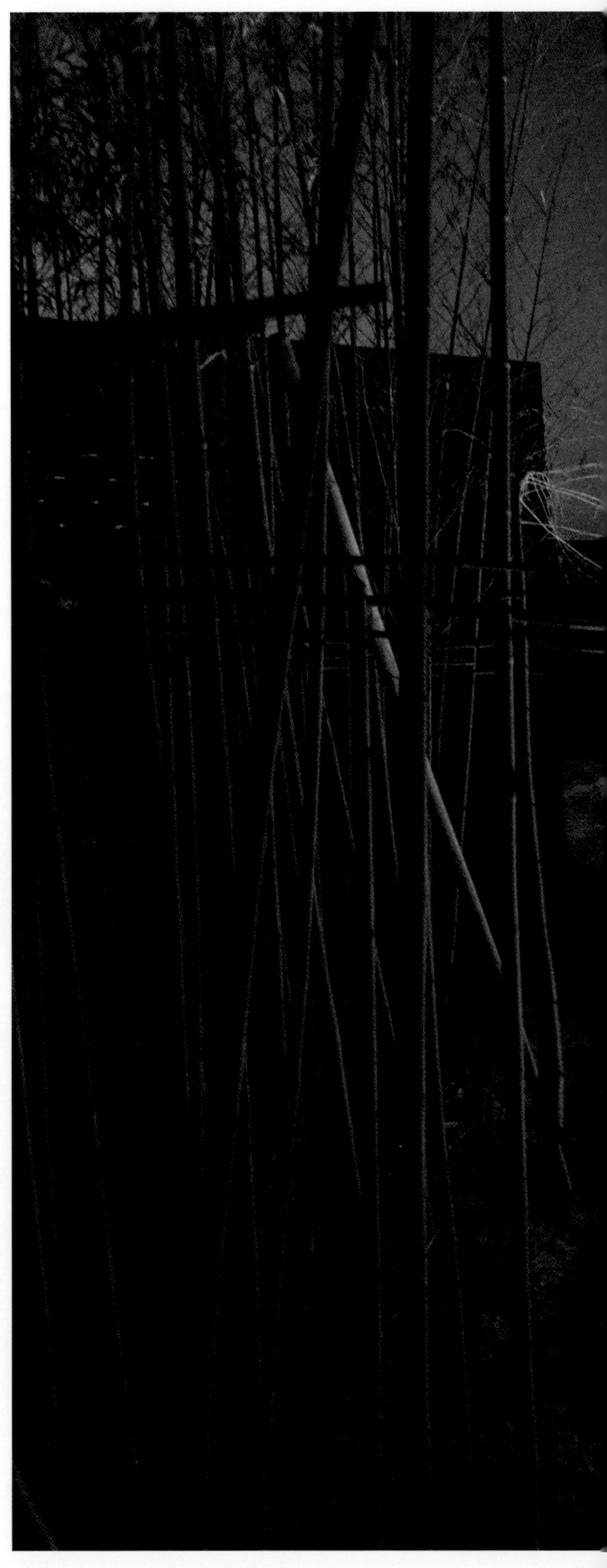

图3-249 成语装置墙
Figure 3-249 A wall featuring Chinese idioms

成语墙
邯郸之名最早出现于
古本竹书记年邯郸地
名之由来现一般以汉
书地理志中三国时魏
国人张宴的注释为源
头邯郸作为国家历史
文化名城其深厚的历
史文化积淀为邯郸留
下了许多的名胜古迹
和历史故事经过千百
年的沉淀凝聚成了现
在的成语典故据中国

4 建成感悟
REFLECTIONS

4.1 前期设计阶段

DESIGN STAGE

项目初期，我院参加了园博园的公开投标，最终呈现出的以现代园林和生态修复为主旨的方案拔得头筹。随后我们组织专业设计团队多次到现场勘查、调研，掌握了翔实的设计资料，并与相关市政单位深入沟通，根据实际情况对方案进行优化完善，增强了方案的可实施性。由于本项目的专业综合度高、技术难度大，在方案深化阶段，设计团队又多次向业主和顾问专家进行专业汇报并与其做技术探讨，从规划到建成历时 21 个月，期间经历了新冠疫情的突发事件，最终于 2020 年 9 月 16 日正式开园。

At the beginning of the project, we won the bidding with a plan themed modern gardens and ecological restoration. Subsequently, our design team conducted multiple site surveys and investigations to gain detailed design information, communicated with municipal departments, optimized the plan based on the input and enhanced the feasibility of the plan. Due to the crossing fields and technical difficulty involved in the project, the design team conducted repeated reports and technical discussions with the owners and consultants in the design development stage. It took 21 months from planning to completion of construction, during which the whole country experienced the pandemic event. Finally, the park officially opened on September 16, 2020.

俞孔坚老师及相关设计团队在项目前期及实施阶段在现场工作的场景。
Photo showing Dr. Yu and the design team conducting onsite works at both design and construction phase.

4.2 施工配合阶段工作心得

REFLECTIONS OF CONSTRUCTION STAGE

4.2.1 建设难点

CONSTRUCTION DIFFICULTIES

园博会项目办会功能需求多，设计施工周期短，如何保质保量地按时完成，施工配合阶段的配合方式尤为重要。

针对园博园项目的专业性、综合性，在施工初期，我院多次组织专业负责人对项目管理方、施工单位进行方案详细介绍和施工图节点解析，避免施工人员对图纸的理解偏差造成施工错误。配合项目工期和专业需求，公司选派不同专业的资深项目负责人对现场进行技术配合、效果把控，并对施工过程中产生的各种问题及时指导到位。同时部分已稳定的方案竖向设计、道路铺装和水系设计快速完善出图，以满足施工工期的需要。在确保设计效果的同时，依据施工时序进行实时施工交底，保证施工人员准确理解设计意图。

作为园林博览会项目，绿化效果尤为重要，因此为了项目更好地落地，我院派驻专业设计师全过程驻场配合。从施工交底开始，创建绿化专项工作群，构建业主以及设计、施工方、管理公司等负责人之间的快速沟通链条，及时解决了施工现场的问题。创建苗木样板质量管理体系，对项目地用到的骨干植物品种统一进行号苗，让所有施工团队在最开始就对施工苗木的选型质量有深刻的认知，做到对绿化方案深刻理解。对施工队负责人及技术员进行施工要点培训，保障施工过程中的各种细节，更好地体现设计理念。因项目交叉施工且各项目专业情况复杂，工期紧张，对场地内的临时变动及施工过程中需要增加的绿化边角区域等，快速提出与周边环境相融合的绿化方案，以保证项目不间断推进。2019 年冬季，配合督导全园 90% 乔木的种植，避免了 2020 年春季疫情带来的诸多不利施工因素，为保证园博会的绿化效果做了充分的准备。多方面多角度配合，主要在开园前约 3 个月预判土建进度对地被绿化效果的影响，快速提出应对开园效果的变通方案，以保障园博园开园效果。

园区建筑在设计周期极其紧张的情况下也必须逐步推进，对于外装饰材料选择、幕墙节点落实、装配工艺设计等，也充分考虑到加工生产与施工安装的周期，所以尽量选择工厂加工、现场安装的方式，以降低施工难度并缩减施工周期。

本项目工程量大，建设周期短，年初疫情又影响了项目工程进度。在施工收尾到正式开园期间，没有时间做相关的试运行来检查各种设备的运行状态及部分土建与景观的边角交接工作情况，给后面正式开园运行带来不确定因素。因此，在未来同类型的项目实施过程中，需要预备时间提前进行湿地水系的通水、试水实验，以确保系统稳定运行。同时，在湿地植物的种植时机和滤料配比方面也应该充分考虑施工期间的日照、温度和降雨等外部环境因素，避免类似问题重复发生。

The Garden Expo has a lot of event holding needs. In order to complete the design and construction of high quality in limited time, the way of cooperation between the design and construction teams during the construction stage is particularly important.

In the early stage of construction, we repeatedly organized detail introduction meetings with the project management and construction teams. Joint analysis was conducted at the construction drawing nodes to avoid construction errors due to misunderstanding of the drawings. We selected senior project leaders of different disciplines to carry out technical cooperation and effect control, and provided timely guidance for various problems during the construction process. At the same time, for the designs that had been established, such as the vertical design, road paving and water system design, we quickly completed the drawings to help meet the construction deadlines. While ensuring the design effect, real-time construction interpretation was carried out according to the construction sequence to ensure accurate understanding of the design intent.

The greening effect is particularly important to a garden expo project. Therefore, we send professional designers to stay onsite throughout the whole construction process, in order to better implement the design. From the beginning of the construction, a special greening work group was created to enable rapid communication between the owner, the design team, the construction team, the management company and other responsible persons. Problems on the site were solved in a timely manner. A seedling model quality management system was created to organize seedling numbering of the backbone plants, so that all construction teams can establish a profound understanding of the greening plan. Construction team leaders and technicians were trained on construction essentials to ensure details in the construction process and better reflect the design concept.

Due to cross–construction and the complexity of the various professions, plus tight construction schedule, temporary changes and green corners that needed to be added during the construction process, greening plan that integrates with the surrounding environment were quickly proposed to ensure uninterrupted project advancement. By supervising 90% of the tree planting in the winter of 2019, many unfavorable construction factors potentially caused by the 2020 spring epidemic have been avoided, and sufficient preparations have been provided for the Garden Expo. About three months before the opening of the park, we predicted the impact of the progress of civil construction on the greening effect, and quickly proposed alternative solutions to ensure the opening of the Garden Expo.

The buildings construction within the park also needed to advance under the extremely tight schedule. For issues such as selection of exterior decoration materials, implementation of curtain wall nodes, and design of installation process, the cycle of processing, production, construction and installation is also fully considered. We therefore tried to choose factory processing and on–site installation method to simplify the construction and shorten the construction period.

The project had a large amount of work with short construction period. Under the impact of the epidemic at the beginning of 2020, there was no time for trial operation to check the status of equipment and the handover of some of the civil works, which had brought uncertainties to the formal opening of the park. In our future projects, it is necessary to make preparations in advance to carry out the testing experiments of wetland water system to ensure stable operation. Meanwhile, in terms of the planting timing and filter material ratio of wetland plants, external environmental factors such as sunshine, temperature and rainfall during the construction period should also be fully considered to avoid recurrence of similar problems.

4.2.2 景观难点（以清渠如许景区为例）

Design Difficulties (Taking Wetland Terraces as Example)

清渠如许节点的设计理念、手法和目标的确定较为顺利，但由于其地形地貌、地质情况、景观功能要求的独特性和复杂性，真正的难点在于中水净化流程设计、台地结构和防水工程设计以及景观细节把控三个方面。

第一，为了保证净化效果，确保处理后水质达标，实现无动力自然布水，人工湿地的形式、形态和水系衔接方式都需要安全、高效和美观。设计团队通过模拟计算发现，由于场地地形和空间限制，常规的人工湿地系统通过“沉淀池 + 表流湿地 + 潜流湿地 + 渗流池”的设置来模拟自然净化湿地的流程无法满足本项目水量和水质的要求。因此清渠如许的水系设计最终以上行垂直潜流湿地取代了原有的渗流池和沉淀池，并采用以管道为主的形式连接所有的净化湿地，在提升水系流通效率的同时避免死水区出现降低净化效果的问题。

第二，传统的人工湿地多采用膨润土防渗毯实现防水要求。清渠如许布置于山地之上，湿地结构多垂直挡墙、衔接部位等，外加场地由建筑垃圾和钢渣堆成，基础强度、沉降情况极不稳定。因此在施工图设计过程中，团队决定改用三布两膜的防水模式，同时解决了防水毯铺设难度大、垂直铺设区域多、容易出现不均匀沉降漏点三大问题，为项目的成功落地和后期安全运行提供了保障。

第三，清渠如许占地面积大、地形复杂，人工湿地和景观栈道交错出现，其水质净化、生态景观和互动参与的综合建设目标有别于传统的、以水质处理为主的单一目标人工湿地。因此除满足净化、安全和通行等基础功能需求外，对细节的处理成为左右项目最终成败的关键。设计团队与实施团队紧密配合，针对常规项目中容易被忽略的细节问题逐一进行深化设计。比如在湿地挡墙外侧毛石砌筑过程中在砂浆层内增加 5% 的防水粉，既解决了挡墙漏水问题，也提升了墙体的美观性；对湿地挡墙跌水口处的设计进行细化，由一个连续切口改为连续复式出水口，提高了水体曝气效率的同时也提升了湿地间衔接处的观赏性；将栈道栏杆改为深灰色，在符合安全规范的前提下，减少栏杆给游人带来的视觉干扰等。

清渠如许节点是生态文明发展与现代景观设计相结合的产物，是城市水体净化和废水资源化利用向“生态化、绿色化、综合化”方向转变道路中的重要尝试。设计把原来的“垃圾山”改造为具有高效中水净化功能、山清水秀的“景观山”，正是邯郸园博园“山水邯郸，绿色复兴”理念的写照。作为园博园中最重要的生态基础设施，清渠如许的设计和实施为未来城市生态景观建设提供了一个宝贵的参考案例。

The design concept, techniques, and goals of the nodes in Wetland Terraces were determined in a smooth manner. However, due to the uniqueness and complexity of the topography, geology, and landscape function requirements, the real difficulty lies in the design of the purification process of the reclaimed water, the terrace structure and the waterproof engineering design, and the control of landscape details.

First, the form, shape and water system connection method of the constructed wetland need to be safe, efficient and beautiful. This is to make sure that the water quality after treatment meets the standard, and realize unpowered natural water landscape. Through simulation, the design team found that due to site topography and space constraints, the conventional wetland system of “sedimentation pond + surface flow wetland + subsurface flow wetland + seepage pond” will not be able to meet the project’s requirements. Therefore, the water system design of Wetland Terraces replaced the original seepage pond and sedimentation pond by upstream vertical subsurface flow wetland, and connected all the purification wetlands in the form of pipelines, which improves the flow efficiency of the water system and avoids the dead water area that reduces the purification.

Second, traditional constructed wetlands mostly use bentonite anti-seepage blankets for the waterproof requirements. Wetland Terraces are placed on the mountain, and the wetland structure is mostly composed of vertical retaining walls, connecting parts, etc., plus the site is made of construction waste and steel slag, which makes the foundation strength and settlement extremely unstable. Therefore, during the construction drawings design, the team decided to switch to the “three-cloth, two-membrane” waterproofing model, which solved the three major problems including difficulty in laying waterproof blankets, too many vertical laying areas, and uneven settlement and leakage.

Third, Wetland Terraces covers a large area with complex terrain, in which the wetlands and landscape boardwalks appear alternately. The construction goals of water purification, ecological landscape and interactive participation are different from traditional single-target artificial water treatment wetlands. Therefore, in addition to satisfying basic functions such as purification, safety and access, the handling of details has become the key to the ultimate success of the project. The design team and the construction team worked closely to intensify the design one by one in response to the details that are easily overlooked in conventional projects. For example, by adding 5% of waterproofing powder in the mortar layer during the rubble masonry on the outer side of the wetland retaining wall, we solved the problem of water leakage and improved the aesthetics of the wall at the same time. The design of the wetland retaining wall's cascade is refined, changing from a continuous cut to a continuous multiple outlet, which increases the aeration efficiency of the water body and improves the appearance of the junction between the wetlands. The boardwalk railing color is changed to dark gray to reduce the visual disturbance caused by railings to visitors, while meeting safety regulations.

Wetland Terraces is the combination product of the development of ecological civilization and modern landscape design. It is an important attempt in the transformation of urban water purification and waste water resource utilization to be ecological, green, and integrated. The design transforms the original "garbage mountain" into a "landscape mountain" with high-efficiency reclaimed water purification function and beautiful scenery, forming a demonstration of Handan Garden Expo Park's concept "Natural Handan, Green Revival". As the most important ecological infrastructure in the park, the design and implementation of Wetland Terraces provide a valuable reference case for the construction of urban ecological landscape in future.

湿土晾晒区

4.2.3 建筑难点（以山水邯郸及涧沟陈展馆为例）

Architecture Difficulties (Taking Handan Pavilion and Jiangou Museum as Example)

山水邯郸位于核心景观区，紧邻水面，是园区最为标志性的主题建筑。由于设计构思来自山水意向，史无前例的超尺度的山形钢架和金属网的运用给设计深化带来极大挑战。在结构层面解决了 60m 的钢结构大跨度问题，首次采用的5万 m^2 户外金属网覆盖面临着安全性、耐久性、艺术性等多方面的考验，我院提出应尽早确定深化生产厂家并进行实地放样与设计确认，但由于种种原因厂家介入过晚，且安装周期过短，导致实际金属网存在整体形态不够顺滑、连接接口粗糙等问题。

涧沟陈展馆位于园博园西南侧涧沟村文旅文化区，属于中型博物馆，建筑主体消隐于地面之下。展陈设计作为陈展馆的重要深化设计单项，方案确定时间较晚，未能与建筑设计密切配合，领会建筑师意图。展陈设计布局过于封闭，建筑内部与庭院未能形成良好的互动关系。建筑本身其实也是体现涧沟文化的一个重要展品，如果能与展陈设计有效结合将会呈现更好的效果。如果后期布展调整的话，希望可以将与庭院互动良好的落地玻璃区域划分出走廊，以走廊串联起各个展厅以及不同的内庭院，从而达到移步换景的效果，弥补方案的遗憾。

Handan Pavilion is located in the core landscape area adjoining the water surface, and is the most iconic building in the park. In order to demonstrate the image of landscape in the design, the use of the unprecedented super-scale mountain-shaped steel frame and metal net brought great challenges. At the structural level, the 60-meter steel structure's large-span problem has been solved. The pioneering use of the 50,000 square meters of outdoor metal net is facing challenges of safety, durability, artistry and other aspects. We proposed to identify the contractor and conduct on-site staking and design finalization as early as possible. However, due to various reasons, the contractor involvement was delayed and the short construction and installation period caused the installed metal net in lack of smoothness in its overall shape and rough connection interface.

Jiangou Museum is located in the cultural area of Jiangou Village on the southwest side of the park. As a medium-sized museum, its main body is hidden under the ground. The exhibition design is an important part of the design of the museum. Due to late plan confirmation, it failed to closely cooperate with the architectural design in order to embody the architect's intention. Therefore, the layout of the exhibition design is too enclosed, which isolates the interaction between the interior of the building and the courtyard. The building itself is actually an important exhibit that reflects the culture of Jiangou. The building would have shown better appearance, if effectively combined with the exhibition design. Should the exhibition layout be adjusted later, it is hoped that the floor-to-ceiling glass area could interact well with the courtyard can be divided into corridors, which connect various exhibition halls and different inner courtyards in series, so as to bring different scenes upon every step made by visitors—as well as make up for our regret of the design.

SUMITOMO

土人设计著作系列

PAST PUBLICATION OF TURENSCAPE DESIGN

为促进中国景观设计学科的研究和发展，北京大学景观设计学研究院与北京土人景观与建筑规划设计研究院同中国建筑工业出版社等出版单位合作，持续出版景观设计理论、实践案例并翻译国外优秀著作。

已经出版的著作包括：

俞孔坚. 理想人居溯本：从非洲草原到桃花源［M］. 北京：北京大学出版社，2020.

YU K J. Ideal landscapes and the deep meaning of Feng-shui: patterns of biological and cultural genes［M］. ORO Editions，2019.

俞孔坚，等. 致力于绿色发展的城乡建设：城市与自然生态［M］. 北京：中国建筑工业出版社，2019.

俞孔坚，牛建宏. 海绵城市十讲［M］. 北京：中国建筑工业出版社，2018.

俞孔坚，张锦，等. 海绵城市景观工程图集［M］. 北京：中国建筑工业出版社，2017.

俞孔坚，等. 海绵城市：理论与实践（上、下）［M］. 北京：中国建筑工业出版社，2016.

AUSTIN G，YU K J. Constructed wetlands and sustainable development［M］. Routledge，2016.

俞孔坚，布莱克韦尔，欧文，等. 北京生态社区：北京市海淀区南沙河区域"反规划"［M］. 北京：中国建筑工业出版社，2016.

俞孔坚，土人设计. 城市绿道规划设计［M］. 南京：江苏凤凰科学技术出版社，2015.

俞孔坚，布莱克韦尔，欧文，等. 浅山区城市发展战略：北京西南部青龙湖案例与启示［M］. 北京：中国建筑工业出版社，2013.

桑德斯. 设计生态学：俞孔坚的景观［M］. 北京：中国建筑工业出版社，2013.

俞孔坚，罗，欧文，等. 台湖展望 中国城乡结合带景观多解规划：北京东南部案例［M］. 北京：中国建筑工业出版社，2012.

俞孔坚，李迪华，李海龙，等. 国土生态安全格局［M］. 北京：中国建筑工业出版社，2012.

俞孔坚，王思思，李迪华. 区域生态安全格局：北京案例［M］. 北京：中国建筑工业出版社，2012.

俞孔坚，李迪华，李海龙，等. 京杭大运河国家遗产与生态廊道［M］. 北京：北京大学出版社，2012.

哈佛大学设计学院. 景观和生态都市主义：北京苏家坨地区多解规划［M］. 北京：中国建筑工业出版社，2010.

俞孔坚，土人设计. 上海世博后滩公园［M］. 北京：中国建筑工业出版社，2010.

俞孔坚. 回到土地［M］. 北京：三联书店，2009.

孔祥伟，李有为. 以土地的名义：俞孔坚与"土人景观"［M］. 北京：三联书店，2009.

斯坦尼茨，弗拉柯斯曼，莫阿林，等. 变化景观的多解规划［M］. 郑冰，等，译. 朱强，等，校. 北京：中国建筑工业出版社，2008.

俞孔坚. 生存的艺术［M］. 北京：中国建筑工业出版社，2008.

YU K J，PADUA M. The art of survival-recovering landscape architecture［M］. Images Publishing Group Ply Ltd.，2006.

北京大学景观设计学研究院，北京土人景观规划设计研究院. 林水城居：土人景观手绘作品集［M］. 长沙：湖南美术出版社，2006.

北京大学景观设计学研究院，北京土人景观规划设计研究院. 数字景观：土人景观表现作品集［M］. 大连：大连理工大学出版社，2006.

俞孔坚，李迪华，刘海龙. "反规划"途径［M］. 北京：中国建筑工业出版社，2005.

俞孔坚，刘向军，李鸿，等. 田：人民景观叙事南北案例［M］. 北京：中国建筑工业出版社，2005.

俞孔坚，王建，黄国平，等. 曼陀罗的世界：藏东乡土景观阅读与城市设计案例［M］. 北京：中国建筑工业出版社，2004.

斯坦纳. 生命的景观［M］. 周年兴，等，译. 北京：中国建筑工业出版社，2004.

俞孔坚，石颖，PUDUA M，等. 人民广场：都江堰广场案例［M］. 北京：中国建筑工业出版社，2004.

俞孔坚，庞伟. 足下文化与野草之美：岐江公园案例［M］. 北京：中国建筑工业出版社，2003.

俞孔坚，李迪华. 景观设计：专业，学科与教育［M］. 北京：中国建筑工业出版社，2003.

俞孔坚，李迪华. 城市景观之路：与市长们交流［M］. 北京：中国建筑工业出版社，2003.

俞孔坚，GAZVODA D，李迪华，等. 多解规划：北京大环案例［M］. 北京：中国建筑工业出版社，2003.

伯恩鲍姆，卡尔森. 美国景观设计先驱［M］. 孟亚丹，等，译. 北京：中国建筑工业出版社，2003.

俞孔坚. 设计时代：国内名设计工作室创意报告［M］. 石家庄：河北美术出版社，2002.

丹尼斯，布朗. 景观设计师便携手册［M］. 刘玉杰，等，译. 北京：中国建筑工业出版社，2002.

俞孔坚. 高科技园区景观设计：从硅谷到中关村［M］. 北京：中国建筑工业出版社，2001.

马库斯，弗朗西斯. 人性场所：城市开放空间设计导则［M］. 俞孔坚，等，译. 北京：中国建筑工业出版社，2001，2016.

西蒙兹，斯塔克. 景观设计学：场地规划与设计手册［M］. 俞孔坚，等，译. 北京：中国建筑工业出版社，2000.

俞孔坚. 景观：文化，生态与感知［M］. 北京：科学出版社，1998，2002，2005.

俞孔坚. 景观：文化，生态与感知［M］. 台北：田园文化出版社，1998.

俞孔坚. 理想景观探源：风水与理想景观的文化意义［M］. 北京：商务印书馆，1998，2002.

俞孔坚. 生物与文化基因上的图式［M］. 台北：田园文化出版社，1998.